# The Structure of Geology

DAVID B. KITTS

SMU PRESS • DALLAS

*Library of Congress Cataloging in Publication Data*

Kitts, David B.
The structury of geology.

Includes bibliographical references and index.
1. Geology—Philosophy. I. Title.
QE6.K57 551 77-7395
ISBN 0-87074-162-4

*To*

Nancy

# Contents

# Preface

THE PAPERS included in this volume were published over a span of thirteen years from 1963 to 1976. Nearly twenty years ago I decided to examine geology in the light of contemporary philosophy of science. I was led to a literature in which the names of Carl G. Hempel, Ernest Nagel, Adolph Grünbaum, and N. R. Hanson figured very importantly. When I began to prepare my first papers for publication I found that these men who had served as such an inspiration through their published works were willing to give their time to encourage and advise me. My debt to Hempel is literally incalculable. His prompt and enthusiastic response to my first requests for advice I later found to be typical of his response to anyone seriously in search of philosophical counsel. My year with Hempel at Princeton stands out as the most intellectually exciting and rewarding of my life.

But my intention was to reach geologists, not philosophers. This has proved to be a challenging and sometimes difficult task. But there were from the beginning those who were willing to help me with it. Foremost among these have been Everett C. Olson and Claude Albritton. Virtually every idea in this book received its first critique from Olson. His remarkable combination of talents have made him an ideal critic and a delightful companion. Claude Albritton has served as a valued adviser and friend not only to me, but to almost everyone else interested in the history and philosophy of geology during the past twenty years. Stephen Gould, whom I first met when he was a graduate student at Columbia, has been from that time an exacting and constructive critic.

I do not recall that G. G. Simpson and I discussed philosophical and methodological issues while I was his graduate student at Columbia, but I must suppose that my interest in these matters was related to, or inspired by, his. In any case his work must serve as a standard by which all work in the methodology of geology and paleontology will be judged in the foreseeable future.

I must also acknowledge the support of my colleagues in geology and the history of science at Oklahoma. They not only tolerated but encouraged my interest in issues that do not clearly fall into either discipline. Prominent among this group are Charles Mankin and Patrick Sutherland in geology and Duane Roller, Thomas Smith, Kenneth Taylor, Sabetai Unguru, and Mary Jo Nye in the history of science. William Livezey made my way easier by giving early administrative sanction to my dual role at Oklahoma.

Whatever my shortcomings as a metageologist, they clearly do not stem from the fact that I have wanted for great teachers and good friends.

Valli Langford has provided invaluable editorial assistance in the preparation of the final manuscript for this book. I am grateful also to Margaret Hartley for her patient and intelligent editing during the book's preparation.

DAVID B. KITTS

*Norman, Oklahoma*
*February, 1977*

# Introduction

GEOLOGISTS AND OTHER SCIENTISTS have not usually found philosophical discussions of the knowledge they have produced to be very interesting. This lack of interest stems in large part, I think, from the fact that scientists come to philosophy with certain expectations; expectations which they have seldom found fulfilled. Philosophers of science do not treat the problems which seem to scientists to be the most "philosophically" interesting. It is not even clear that what philosophers of science are doing can even be seen as a proper part of philosophy. It seems to many scientists that the questions which philosophers of science are in the habit of asking are, if not "scientific" questions, at least questions which are asked by scientists and to which they give more or less satisfactory answers. Philosophers of science have apparently not succeeded in conveying just what special knowledge or insight they have to offer. I do not intend to undertake the difficult task of bridging the gap between science and philosophy. I do hope to make clear, however, what it is I have attempted to accomplish wholly within, but by no means exhausting, that rich analytical apparatus provided by contemporary philosophy of science.

When geologists see the words *geology* and *philosophy* in juxtaposition they are likely to think of those great debates of the nineteenth century which were concerned with the bearing of new knowledge about the history of the earth upon an extended and ultimate world view. Philosophically inclined geologists often expect twentieth-century philosophy of science to shed some new light upon this old problem. Is it not an important, perhaps the most important, task of philosophy to tell us what the world is *really* like, or at least to suggest ways of finding out? It would be a mistake to suppose that many contemporary philosophers of science would answer this question in the affirmative. We must proceed with caution in characterizing so broad and diverse an activity as philosophy of science, but it is important to recognize that contemporary Anglo-American philosophy of science grew out of a school which had explicitly repudiated what was considered to be an excessive concern with metaphysics in traditional philosophy. It has been my intention to avoid metaphysical questions most scrupulously. One very good reason for this is that I am not a metaphysician. But there is a more immediate and practical reason. I am convinced that the failure to produce a coherent and illuminating account of the structure of geological knowledge in the nearly two centuries since the publication of Hutton's *Theory of the Earth* has resulted from a persistent confusion between purely metaphysical questions on the one hand and other legitimate philosophical questions on the other. The confusion is clear in those nineteenth-century debates to which I have alluded. The question of whether nature was "really" uniform was virtually never distinguished from the question of the inferential consequences of supposing that the principles of contemporary science could be invoked without temporal restriction. It is not my intention to fault our intellectual forebears for failing to make a distinction which I consider to be a legitimate one. I shall attempt to demonstrate, however, that it is not only possible but instructive to examine such important issues as the role of physical theories in the systematization of geological knowledge without considering, or even surreptitiously presupposing, anything about the ontological status of the theoretical entities to which these theories refer.

Having failed to find the answers to ontological questions, or even much concern for ontological issues, in philosophy of science, a scientist may still pursue the subject in the hope of finding some instruction in "scientific method." But again he will, I think, be disappointed. When a scientist talks about "scientific method" he almost always has in mind a procedure by which scientific knowledge is discovered. Despite the occasional insistence that scientific discovery is a proper subject for philosophical inquiry (see especially Hanson, among others) the principal thrust in contemporary philosophy of science has been aimed at an elucidation of the procedure by which scientific knowledge is justified. This is not to say that philosophers necessarily hold discovery to be some mysterious process beyond the reach of rational analysis. It is only to say that the inferential schemes which they have in fact formulated relate much more obviously to the justification of knowledge than to its discovery. In the papers contained in this volume I have attempted to separate clearly the "context of discovery" from the "context of justification," and I have had a great deal more to say about the latter than the former.

It may appear to a geologist that a discipline which has little to say about how we make scientific discoveries, and no more to say about the ontological status of the things discovered, is hardly worth pursuing in a quest for philosophical illumination. I have attempted to show in these papers that even after the removal of questions about discovery and "reality" there remains an important class of questions that can be asked about scientific knowledge. Contemporary philosophy provides a means of asking these remaining questions and sometimes provides a means of answering them.

It is easier to say what I have not done than to describe what I have done. But whatever operations I have performed, I have performed upon geology. Colloquial usage notwithstanding, geology and the earth are not the same thing. The study of geology is a study of a body of knowledge rather than a study of concrete entities such as rocks. It is the relationships among the various parts of geology that I mean to get at when I talk about the *structure* of geology. The parts of a body of knowledge are themselves items of knowledge which must be embodied in statements. The analysis of the structure of a body of knowledge will proceed largely in terms of a system which is suited to a treatment of relationships among statements. Such a system is, of course, logic. But geology, being a natural rather than a purely formal science, presents the further problem of the relationship between statements and the features of the world which they are meant to describe. The study of the structure of science, an inquiry which is concerned with the way in which statements may be justified empirically and in terms of one another, constitutes a major part of that branch of philosophy of science often called the "theory of science."

Just as science employs terms designed to enhance our understanding of the world, so a theory of science employs terms designed to enhance our understanding of science. And if a theory of science is to provide an understanding of science different from or beyond that which can be provided by science itself, it must invoke terms which do not figure significantly in scientific discourse. It is true that terms such as *law, theory, induction, verification,* and *proof* may figure in science, but they are likely to have different and more precise meanings when they occur in philosophical talk *about* science. An important feature of these philosophical terms is that they are purported to have applicability to all sciences. Because of this a theory which invokes them may be able to accomplish what is thought to be a particularly important task: to compare, and perhaps even in some sense to unify, the various branches of science.

Unfortunately, a distinction between a science and a theory of science which rests upon a distinction between statements about the natural world and statements about statements about the natural world presents some difficulties. One of these is that disagreements concerning broader philosophical matters may lead to disagreements about whether a given statement falls into one or the other of these categories. It is, furthermore, true that the division of labor between science and philosophy is not as clear as the foregoing remarks might imply.

The distinction between *science* and *theory of science* is easier to maintain in geology than it is, for example, in physics. Geologists have tended to see themselves as dealing with more or less obvious matters of fact which they encounter not in the contrived and circumscribed theater of the laboratory, but in the broader and richer reality of the "field." It can easily be shown, I think, that the geological tradition has fostered from the beginning a radically empirical and therefore, in the minds of many geologists, a notably untheoretical geology. This view is sustained by the undeniable fact that the primary preoccupation of geologists is with concrete things like mountains and rivers rather than with presumptive entities like electrons and protons.

In formulating special relativity theory, Einstein is compelled to confront problems which might as well be called epistemological as scientific. The geologists never sees himself in such a situation. He sees himself as confronting events which differ from the events of the world in which he lives only in being removed in time. The vocabulary he uses to describe those temporally distant events does not differ from the vocabulary he uses to describe the events of his own time. For this reason he is not continually reminded that the events of the historical past are not got at in the same way as the events of the present. This proper and understandable preoccupation with real events has apparently led most of those who have attempted to characterize geology explicitly to underestimate badly the theoretical content of the science. It has led some to regard geology as a protoscience waiting for its great synthesizer. But the Newton of geology has been here and gone. His name was Isaac Newton. The intimate connections between classical mechanics and other physical theories and the events which they comprehend can be delineated within science; and to the extent that physical theory is legitimately and rigorously employed in geological inference, this task is being accomplished day by day. The instrument of philosophical analysis does not provide a means of doing that task better or doing it over again. Philosophical analysis does not repeat science. It follows science in the hope of illuminating it. The illumination I seek here is an explicit delineation of the theoretical structure of geology. But I have claimed that geology requires a special treatment because it is, in an important sense, a special science.

Geologists are sometimes heard to remark that geology is historical. Surprisingly little methodological capital is made of this recognition. As a matter of fact, when geologists discuss the fundamental structure of their science they are likely to do what nearly everyone else does, and that is to turn to physics as a standard for comparison. A striking result of this questionable procedure is that prediction, which is supposed by geologists to play a critical role in physics, is often supposed by them to play an equally critical role in geology. Geology may then be faulted as a "protoscience" for its lack of predictive efficacy.

A moment's reflection will show that geology need not fail to yield predictions that are comparable in precision and circumspection to the predictions yielded by physics. But geologists are simply not interested in events that are so simple and limited in spatial and temporal extension as those which can be predicted. Geologists, if they predict at all, want to predict events that are comparable in complexity and extension to the events they are able to retrodict. Geologists are, in short, preoccupied with what they regard as historically significant. The concept of historical significance is nebulous and is likely to remain so. But it is clear that in the inference of historically significant events, whether it be predictive or retrodictive, the emphasis will be placed upon the detailed ways in which these events differ from one another, rather than upon the fundamental ways in which they resemble one another. Geologists usually attempt to get at this feature of historical science by pointing to the "uniqueness" of historical events.

It may seem that in physics the goal is not to infer complex events but, on the contrary, to infer the simplest events possible. There is a very good reason for this. Physicists, unlike geologists, are not interested in events for their own sake. They are interested in events as a means of getting at theories. It is hardly ever a *critical* feature of a description of an event in physics to note its particular spatial and temporal location. It is always critical to locate a historical event in some particular region of space and time. In geology *locating* an event is often an important means of establishing its "uniqueness." Geologists can predict as well as physicists can. But prediction, for whatever reason, seldom yields events which are interesting for their own sake. If geologists were to ask of themselves only what physicists ask of themselves, there would be no justification for faulting geology on the ground that it failed to yield predictions. As a matter of fact geologists want, and even expect, a great deal more from prediction than physicists do. The current concern with the prediction of earthquakes provides dramatic confirmation of this.

It is the preoccupation with events, and with the singular statements which describe them, that makes geology historical. Popper has said (1957, p. 143), "The situation is simply this: while the theoretical sciences are mainly interested in finding and testing universal laws, the historical sciences take all kinds of universal laws for granted and are mainly interested in finding and testing singular statements." A geologist will not object to the claim that geology is a historical science. He may object, however, if this assertion is taken to mean that it is a discipline largely or wholly concerned with the derivation and testing of singular descriptive statements. Contemporary geologists have sought to counter the view that geology is "merely descriptive." But to say that geology is primarily directed toward generating singular descriptive statements cannot, of course, be taken to mean that geologists are largely concerned with describing things, if describing is taken to mean simply reporting what our senses tell us. Leaving aside, for the moment, the question of whether or not our senses alone can *ever* provide the basis for description, it is clear that the most interesting and significant geological descriptions are of events that lie far beyond the range of our senses. In the papers contained in this volume I have stressed the complex inferential context in which descriptions of the remote past are derived and tested. This inferential context is theoretical either in employing notions directly from the theoretical sciences, or in employing geological generalizations which have significant logical properties in common with, and important conceptual relationships with, formal theories. To say that historical science is wholly concerned with singular descriptive statements is as mistaken as to say that theoretical science is wholly concerned with theories. The context within which historical scientists test singular descriptive statements is theoretical. The context within which theoretical scientists test theories is singular descriptive.

The difference between the theoretical sciences and the historical sciences does not lie in the theories which are invoked or in the inferential use to which these theories are put. It lies rather in what those engaged in the two kinds of sciences see as their goal. For historical scientists, singular descriptive statements are the end and theories are a means to that end. For theoretical scientists, theories are the end and singular descriptive statements are a means to that end.

Some geologists have taken my remarks about the historical nature of geology to be prescriptive rather than descriptive. When I say that geology is historical I mean to claim that geologists see as their principal goal the derivation and testing of singular descriptive statements. This entails no judgment that geology must be as I have described it, or that it should be.

A science which took the planet earth as its subject matter *might* be directed toward the development of a set of general laws about the earth and its parts. In such a theoretical earth science, statements about events would not be important for their own sake but would be seen as providing instances of verification and falsification for the laws. Another earth science, conceivable in principle, would take history to be a chronicle of the changing character of spatially universal laws rather than a chronicle of changing events under immutable laws. In an extreme version of this science, assumptions might be made about particular events so that the validity of laws could be tested for various times in the past.

Geologists have claimed from time to time that geology contains elements of both of these hypothetical earth sciences. I have denied this claim for contemporary geology. Contemporary geological practice is radically historical in just the sense that I have claimed. An examination of the geological literature will, I am quite certain, support this contention.

I have argued further that these hypothetical earth sciences raise methodological problems that have been given scant attention, largely because a solution would rest upon a distinction that geologists almost never make: a distinction between laws and theories on the one hand and descriptions of events on the other.

Hutton, being a self-conscious and sophisticated empiricist, regarded the events which we encounter in our own time as a means of discovering the "natural order." This natural order he assumed to hold for all time and thereby to impose a limit upon what we may suppose to have been possible in the past. Others, and I believe Lyell to have been one of these, held that events, rather than being the means of discovering a universal order, themselves imposed the order. This view amounts to the dictum that all that can be is. There is nothing wrong with a theory which specifically stipulates limits upon a range of events—for example, upon an upper limit for some rate. Relativity theory does this for very good theoretical reasons. The difficulty that has plagued geology is that many have sought to impose a limit upon events that is more restrictive than the limit imposed by the physical theories which everyone agreed comprehended the very events in question.

Dramatic recent advances in geology have been possible because the limit imposed upon what we may suppose to have occurred in history has been provided by the most comprehensive physical theories rather than by the events which we can observe. But the discussion of geology has lagged far behind the practice of geology. This is evident in almost all discussions of geological methodology, and particularly in that vast literature devoted to earth science education. In a passage from the teacher's guide to *Investigating the Earth*, it is stated (Earth Science Curriculum Project, 1967, p. 3):

> The body of scientific knowledge at any given moment represents only one stage in man's effort to understand and explain the universe. Today's useful theories may be the half-truths of tomorrow. In this investigative approach, science is presented as inquiry, as a search for new and more accurate knowledge about the earth. The student learns through experiences in the laboratory by using the scientific methods that have led to our present knowledge of science, as well as to a feeling of incompleteness and uncertainty of this knowledge.

A science teacher who has been subjected to a curriculum that presented science as a static body of established truths is likely to respond enthusiastically to such a program. But however attractive the program may be, it raises certain difficulties that must be recognized. A theory of science teaching that is directed toward revealing the means by which scientific knowledge is acquired necessarily presupposes a theory of scientific knowledge. A serious defect of the investigative theory of teaching is that the theory of knowledge that underlies it has never been critically examined or even explicitly stated. The authors of *Investigating the Earth* apparently felt that the "methods that have led to our present knowledge of science" are so well understood and agreed upon that their explicit treatment is unnecessary. In the absence of such a treatment, we can only infer the philosophical assumptions underlying the investigative approach. The only theory of knowledge fully consistent with the investigative theory of teaching is a particularly naïve form of inductivism. Inductivism holds that the acquisition of scientific knowledge begins with observations and proceeds to a knowledge of principles and theories. This is a venerable doctrine that has been with us in one form or another since the beginning of science. Many versions of inductivism, including the one held by most teachers of geology, rest upon the assumption that observations and the descriptive terms in which they are reported have a meaning that is independent of any general preconceptions. Thus, observation provides a means of approaching reality with an "open mind." But geological observation and geological generalization take place almost wholly within a complex system of general preconceptions—a system so complex that we cannot hope with any reasonable effort to identify all of its components. There is, however, a readily recognizable part of this elaborate system. It is the fundamental principles of contemporary science that are largely contained in physical, chemical, and biological theory. These theories provide geology with an inferential apparatus of great power and demonstrated utility. If geologists are to employ this apparatus, their descriptions and generalizations must be formulated in terms consistent with a vast body of knowledge that is assumed to be, at least for the purposes of geological inference, unproblematic. In a very significant sense, then, geologists do not approach their subject matter with an open mind. They do not give equal weight to what their senses tell them. They take into account only that which is already imbued with theoretical significance, and they do not formulate principles and generalizations by an inductive enumeration of observations.

The investigative theory of geology teaching fails to emphasize the critical dependence of geological observation and inference upon an unproblematic background of theoretical presuppositions. This failure is sustained, in my judgment, by a mistaken assumption about the nature of scientific, and particularly geological, knowledge. It is indeed true that the study of geology is more interesting and meaningful if emphasis is placed upon the means by which geological knowledge is acquired. The acquisition of geological knowledge depends critically upon the presupposition of general notions, the most obvious among which are the theories of chemistry and physics. Geologists do not approach their subject matter with an open mind, and they should not expect their students to do so. Theoretical preconceptions should assume at least as much significance in the teaching as in the practice of geology.

## References Cited

Earth Science Curriculum Project, 1967, Investigating the earth, teachers' guide, Part 1: Boston, Houghton Mifflin Company, 446 p.

Popper, K. R., The poverty of historicism: New York, Harper and Row, 166 p.

# The Structure of Geology

CHAPTER ONE

# Historical Explanation in Geology

## INTRODUCTION

Geologists have long maintained that their discipline differs strikingly from chemistry and physics in being historical. There has been almost no explicit consideration by geologists, however, of the particular methodological problems imposed by the historical dimension. Historians and philosophers, on the other hand, have considered the problems of historical method at great length, and in recent years there has been, it seems, an increased interest in the methodology and philosophy of history. It is my purpose here to consider the extent to which the issues raised in recent discussions of explanation in human history are pertinent to the problem of explanation in historical geology. Because paleontology presents special methodological problems that I shall later discuss in detail (see pp. 148-70), I have omitted at this point any consideration of geological explanations containing paleontological concepts.

It would be a serious mistake to attempt to apply the principles of historical method to geology simply because there is general agreement among geologists that their discipline is in some sense historical. It will be necessary to consider carefully the ways in which the study of human history is like the study of geology and the ways in which it is different.

There is, to begin with, a very obvious difference in the subject matter of the two fields. Historians with whom I have discussed historical method have repeatedly cautioned me—with good reason, I think—not to ignore the methodological consequences of this fundamental difference. The difference in subject matter between history and geology does indeed lead to methodological consequences, and some of these consequences are quite obvious. For example, it is clear that one would be unlikely to find any methodological principle that might be appropriately applied to historical geology in the works of Collingwood and others of the "idealist" school of historical interpretation who believe that "thoughts" constitute the proper subject matter of history. Similarly, explanation by "colligating events under appropriate conceptions" as outlined by Walsh (1951) could not be meaningfully related to geology because, as Walsh points out, this form of explanation depends upon the special subject matter of the historian and often proceeds in teleological terms.

The most obvious way in which geology and history are alike is in the place of time in the two systems. It certainly cannot be said, as Ellison does (1960, p. 982), that "Geology is the only discipline that stresses the importance of time," for there is no scientific discipline in which time does not play an essential role and in which events of a later time are not explained in terms of events of an earlier time. History and geology may be distinguished from other disciplines in the magnitude of the time span involved; but this difference in degree does not, it seems to me, involve any special logical problems, although it almost always results in special empirical problems.

Later in the paper quoted above, Ellison (p. 982) implies what I believe to be the fundamental difference between history and geology, and large parts of biology and astronomy, on the one hand and the nonhistorical natural sciences on the other when he states: "Placing of geological events in their proper time position and perspective in the interpretation of earth history represents the goal of much of geological thinking." Geologists and historians are interested in determining the specific dates and the specific places at which events took place and in constructing a historical chronicle on the basis of these determinations. This point has been repeatedly made for history, and in fact Popper (1957, p. 143) has used this characteristic as the basis for distinguishing two distinct kinds of scientific activity: "The situation is simply this: while the theoretical sciences are mainly interested in finding and testing universal laws, the historical sciences take all kinds of universal laws for granted and are mainly interested in finding and testing singular statements." Cohen expresses the same point of view (1947, p. 38-39) in the statement:

> Those who insist that history is a science in the same way in which physics is a science often mean to assert that the subject matter of history is not the individual events but the laws or repeatable patterns of human behavior. Those, however, who do so obviously confuse history with sociology. A science of sociology would be concerned with general laws and would leave to history the consideration of what actually happened in definite places at given times.

The science of geology as a whole, not just that part of it called "historical geology," has been primarily directed toward the goal of "finding and testing singular statements." Geologists may leave to others the formulation of physical and chemical laws that they employ as a means of inferring what happened at a particular time and place in the past in the same way in which a historian may leave to others the formulation and testing of many of the laws that he employs. On the other hand, most geologists, and I am sure most historians, would object to being characterized as individuals who "take all kinds of universal laws for granted." Geologists have made liberal use of physical and chemical laws, but they have found that these laws alone are not sufficient for the task they have set for themselves. They have consequently gone on to formulate, if not laws, at least generalizations, which they employ in their inferential operations. On the basis of an examination of a great deal of geological literature and of conversations with many geologists, however, I have come to the conclusion that even among those geologists who are mainly concerned with the attempt to originate new generalizations, these generalizations are regarded as means to an end rather than as ends in themselves, the end being the construction of a chronicle of specific events occurring at specific times. With all the emphasis in recent years upon "earth science" and the theoretical physical-chemical foundation of geology, the primary concern with specific events still dominates our discipline. Melton (1947, p. 52) emphasized this point when he stated: "It [geology] is not at all concerned with the confirmation or denial of any theory; but it is deeply concerned with the accurate recital of the events of the earth's history."

So important do I regard this concern with specific events that I have confined my discussion here to the explanations of specific events.

It is the concern with the specific from which the assertions about the complexity of geology stem. Geologists are only secondarily interested in ideal situations that can be produced in a laboratory and are mainly interested in the sometimes overwhelming complexity of "what actually happened." Experimentation has had a relatively unimportant place in geology, not only because geological experiments entail difficult technical problems, but because, if we view experimentation as a means of testing generalizations, geologists have not been very interested in the kind of knowledge that it can yield. Experimentation is playing an increasingly important role in geology, and it is directed, as in any other natural science, toward the testing of generalizations, generalizations that are in turn regarded as instruments which allow the "reconstruction of the past."

Thus, notwithstanding the very fundamental differences in subject matter, there are two important points of similarity between the study of human history and the study of geology. The first point of similarity, that both disciplines deal in a vastly extended time dimension, results in common empirical problems. The second point of similarity, that in both disciplines the basic concern is with deriving and testing of singular statements, results in, if not common logical problems, at least a common logical point of view.

## DEDUCTIVE EXPLANATION IN SCIENCE

Much discussion of historical explanation revolves about the question of the resemblance, or lack of resemblance, between explanation in history and explanation, particularly deductive explanation, in science. It is therefore convenient to consider deductive scientific explanation before moving on to a discussion of historical explanation.

The idea of explanation as a deductive operation goes back to Aristotle. Throughout subsequent history there have been discussions of the adequacy of Aristotle's account of a deductive scientific explanation, and there have been various proposals for modifications of the Aristotelian model. It is upon recent characterizations of the deductive scientific explanation, particularly those of Popper (1957, 1959), Hempel and Oppenheim (1948), and Nagel (1961), that the discussion that follows is based. I am neither qualified nor inclined to discuss technical problems of the logic of explanation, and I have consequently made an effort to avoid them.

Popper (1957, p. 122) describes the characteristics of a causal or deductive explanation of a particular event in the following words: "I suggest that to give a causal explanation of a certain *specific event* means deducing a statement describing this event from two kinds of premises: from some *universal laws*, and from some singular or specific statements which may we call the *specific initial conditions*." In most modern discussions of explanation the statement describing the event to be explained—that is, the conclusion—is called the "explicandum," while the singular statements and universal laws necessary for the deduction of the explicandum, that is, the premises, are called the "explicans."

In a lucid analysis of explanation Nagel (1961, p. 32) recognizes two logical requirements for adequacy for a deductive scientific explanation of the occurrence of some event or the possession of some property, as follows: "The premises must contain at least one universal law, whose inclusion in the premises is essential for the deduction of the explicandum. And the premises must also contain a suitable number of initial conditions." The first requirement excludes from the category of explanations deductions containing only singular statements, that is, statements concerning particular objects, times, and places. There are few scientists who would accord the status of explanation to this kind of deduction. The second condition is necessary because general laws will yield a conclusion in the form of an unconditional singular statement only when combined with statements about initial conditions which themselves have the form of singular statements.

Nagel in a discussion of the "epistemic" conditions of adequacy (ibid., p. 43) rejects the Aristotelian requirement that the premises must be true and must be known to be true as "much too strong," since an acceptance of this condition would entail the rejection of most of the explanations offered in science. Because some empirical restriction of the statements admitted to the premises is necessary in order to exclude explanations containing *ad hoc* statements, Nagel suggests (p. 43):

> A reasonable candidate for such a weaker condition is the requirement that the explanatory premises be compatible with established empirical facts and be in addition "adequately supported" (or made "probable") by evidence based on data other than the observational data upon which the acceptance of the explicandum is based.

The second condition is extremely important for our later consideration.

For Hempel and Oppenheim (1948) the empirical condition of adequacy for a deductive explanation requires that the sentences constituting the premise be true. They point out that to substitute for the requirement of truth the requirement that the premise be highly confirmed by all relevant evidence available at a particular time might lead one to the awkward position of asserting that an explanation was correct in the past and that in the light of subsequent empirical finding it ceased to be correct. Whatever the formal requirements it is clear that we do consider an explanation "correct" or adequate if the premise is in fact highly confirmed by all relevant evidence available at the time the explanation is proposed.

A point that has been repeatedly emphasized by recent authors is the logical similarity between explanation and prediction. For example, according to Hempel and Oppenheim (1948, p. 138), "the same formal analysis, including the four necessary conditions, applies to scientific prediction as well as to explanation," and according to Popper (1957, p. 124), "the use of a theory for the purpose of *predicting* some specific event is just another aspect of its use for the purpose of *explaining* such an event." This similarity can be seen if we consider first the case in which a specific event described in the explicandum has occurred and a suitable group of statements is provided for the explicans. We say in this case that the event has been explained. If, on the other hand, the statements of the explicans are provided before the occurrence of an event and a statement describing the event has been deduced from the explicans, we then say that a prediction has been made. Finally, it may be possible to deduce a statement about some past event from a statement describing some present condition together with a suitable law, in which case we may say that a "retrodiction" or "postdiction" has been performed. Hempel (1958, p. 38) has applied the term "deductive systemization" to the three operations described above. As Nagel has said (1961, p. 32, n. 4), "inquiries may differ according as they are directed toward finding and establishing one type of premise rather than another." He goes on to point out that first "we may note the occurrence of some event, and then seek to explain it by discovering some other event which, on the basis of an already established law, is assumed to be the condition for the occurrence of the given event." And second, "we may note the occurrence of two or more events, suspect them to be significantly related, and attempt to discover the laws which formulate the specific modes of dependence between events of that character." And finally, "in the attempt to explain some events, inquiring may be directed to discovering both types of suitable explanatory premises."

Hempel and Oppenheim (1948, p. 138) summarize the characteristics of a deductive explanation in the diagram shown in figure 1.

It is clear that an understanding of the nature of deductive explanation in science must rest upon an explicit characterization of the concept "natural law." Later I shall discuss in detail the subject of geologic laws and generalizations. For this reason I shall not treat the subject at any length here. There has been a great deal of discussion of the problem of laws in the literature of the method and philosophy of science, and there are a number of points about which no general agreement has been reached. This lack of agreement is not surprising since, as Nagel (1961) has pointed out, "scientific law" is not a term that has been rigidly defined in each of the sciences. It will be sufficient for my purposes to point out that a law employed in an operation, whether explanatory, predictive, or retrodictive, which yields a deductive conclusion must be of *strictly universal form*. As Hempel has said (1958, p. 39), "A statement of this kind is an assertion—which may be true or false—to the effect that all cases which meet certain specified conditions will unexceptionally have such and such further characteristics."

Many useful and important generalizations that are employed in scientific systematization are not of strictly universal form but rather of statistical form. To quote Hempel again (p. 39), "A statement of this kind is an assertion—which may be true or false—to

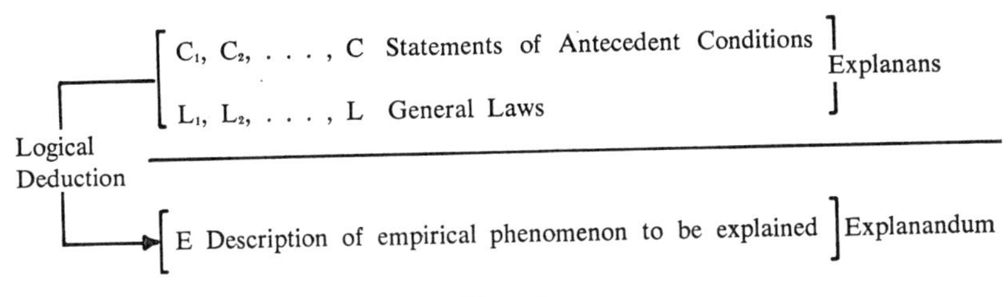


FIG. 1

the effect that for cases which meet conditions of a specified kind, the probability of having such and such further characteristics is so-and-so much." Thus in arguments that employ statistical generalizations, whether they be explanatory, predictive, or retrodictive, the conclusion does not follow deductively from the premises. For this reason Hempel has applied the term "inductive systematization" to inferential procedures involving statistical generalizations.

Scriven (1958, p. 193) has gone so far as to say, "It is virtually impossible, as previously remarked, to find a single example of something that is normally called a law in science which can be precisely formulated in non-probability terms," and further (p. 193), "I think it certain that all scientific explanations, with quite negligible exceptions, are 'statistical' or better, probabilistic; and the authors of the deductive model explicitly exclude consideration of such."

## THE COVERING-LAW MODEL OF HISTORICAL EXPLANATION

In 1942 Hempel published a paper entitled "The Function of General Laws in History." This work has been tremendously influential. Most of the papers written on historical explanation during the past twenty years have either supported or attacked Hempel's position as expressed in this work.

Hempel begins with a characterization of scientific deductive explanation which is essentially like that which he and Oppenheim were to give in more detail in their 1948 paper on the logic of explanation. His position regarding explanation in history is presented in the following quotation (p. 39):

> The preceding considerations apply to *explanation in history* as well as in any other branch of empirical science. Historical explanation, too,

aims at showing that the event in question was not "a matter of chance," but was to be expected in view of certain antecedent or simultaneous conditions. The expectation referred to is not a prophecy or divination, but rational scientific anticipation which rests on the assumption of general laws.

Hempel goes on to say (p. 40) that most explanations offered in history, "fail to include an explicit statement of the general regularities which they presuppose." This feature of historical explanations is accounted for by the fact that many of the regularities in question concern individual or social psychology with which everyone is supposed to be familiar, and furthermore that "it would often be very difficult to formulate the underlying assumptions explicitly with sufficient precision and at the same time in such a way that they are in agreement with all the relevant empirical evidence available."

Hempel's view of historical explanations is summed up in the following passage (p. 42):

> What the explanatory analysis of historical events offer is, then, in most cases not an explanation in one of the meanings developed above, but something that might be called an *explanation sketch*. Such a sketch consists of a more or less vague indication of the laws and initial conditions considered as relevant, and it needs "filling out" in order to turn into a full-fledged explanation. This filling-out requires further empirical research, for which the sketch suggests the direction.

We may conclude, then, that in Hempel's opinion many historical explanations have the general character of explanations in the sciences but fail to satisfy the rigid conditions of adequacy enumerated in the discussion of the deductive explanation presented above. We should note, however, as Hempel does, that this failure to satisfy conditions of adequacy is not limited to historical explanations but is characteristic of explanations offered in other fields as well. The model of historical explanation proposed by Hempel has come to be known as the "covering law" or "deductive" model of historical explanation. Among those who generally subscribe to this view are Popper (1957), Nagel (1952, 1961), and Gardiner (1952).

Hempel notes that many historical explanations seem to be statistical in character, that is, that they seem to be based on the assumption of a probability hypothesis rather than on a deterministic law that can be expressed as a universal conditional. Nagel (1961, p. 561) in consideration of the same point states:

> The incompleteness of the premises when measured by the standards of valid deductive reasoning and their formation of necessary rather than

sufficient conditions for the occurrence of events, are two generally acknowledged traits that explicate in part the sense in which historical explanations are "probabilistic."

In other words, the premises of a probabilistic explanation are not a sufficient basis for the prediction of the event explained; and in fact, as Nagel puts it, "the truth of the premises in an historical explanation is entirely compatible with the falsity of its conclusions."

A number of other authors have referred to the probabilistic character of historical generalizations and explanations. For example, Frankel (1957, p. 144) has said, "Most historical generalizations do not seem attempts to state invariant relations, but only correlations of significant frequency." Scriven's (1959) criticism of the application of the deductive model to historical explanation is, at least in part, based upon the probabilistic character of the "truisms" in terms of which, he maintains, most events in human history are explained. Furthermore, Scriven objects to characterizing as "incomplete" explanations in which no generalizations or laws are stated because, in his opinion, laws should not be regarded as a part of an explanation, but rather should be regarded as a part of the *justification* for an explanation.

Gallie (1955) maintains that "genetic" explanations are particularly characteristic of historical discourse. For Gallie a characteristically genetic explanation consists of the indication of some kind of continuity between a *series* of temporarily prior and subsequent events. Presumably this continuity is inferred in terms of some "probabilistic" generalization. Gallie emphasizes that the genetic explanation does not have predictive power, that is, that the prior conditions together with whatever generalizations are employed do not constitute a sufficient basis for prediction of the subsequent event. In Nagel's (1961) view a genetic explanation is analyzable into a series of probabilistic historical explanations.

Most discussions of explanation in history take "past events" as starting points and proceed immediately to a consideration of how, or in what sense, the historian explains these events. Philosophers have from time to time felt it necessary to point out that the propositions with which historical inferences begin are not themselves historical but are assertions about what exists now. Dewey (1938, p. 232-33), among others, has made this obvious point because he felt that in losing sight of it historians too often failed to make explicit the inferential procedure by which statements about what now exists could serve as a justification for statements about past conditions or events. It is not surprising that historians are rarely explicit about this procedure, for nothing is seemingly more trivial than an explication of the method by which a historian infers, for example, the occurrence of a past event from a document purported to have been written by someone who witnessed the event; seemingly trivial because most of the generalizations that are employed in such a procedure turn out to be perfectly familiar assertions about human behavior or human "nature."

I do not mean to imply that this very important question has been generally ignored. It has not. For example, Ayer (1958, p. 182-83) has this to say about the process:

> Since no event intrinsically points beyond itself, our reason for linking a later with an earlier event, for assuming that the one would not in the given circumstances have occurred unless the other had preceded it, must lie in our acceptance of some general hypothesis; that is to say, we account for the later event by correlating it with the earlier.

Gallie (1955, p. 178) observes: "When a historian or epigraphist interprets a document he always looks, *inter alia*, for a characteristically historical explanation of it." He goes on to make a clear distinction between an explanation of the document, the prior operation, and an explanation of an event which the document purports to describe. And finally Renier (1950, p. 157-58), apparently in agreement with the above views, states:

> The mind which contemplates these traces sets to work at once. It ceases to look upon a trace as being merely the end term of a sequence of occurrences. It gets hold of the trace and makes it the starting point for an inference. For the trace can be accepted, by anyone who knows how efficaciously experimental science makes use of its data, as an event connected by a "natural law" with another event earlier than itself.

## HISTORICAL EXPLANATION IN GEOLOGY

### *Primary Historical Inference*

It is, I believe, logically acceptable and fruitful to view primary historical inference in geology as involving the formulation of a probabilistic historical, and usually genetic, explanation of geologic "documents." Few geologists will object to the use of the term "explanation" in this connection, since in most cases they explicitly employ this term to designate the operation of primary historical inference, although the term "interpretation" is frequently used. The operation must be regarded as fundamentally probabilistic simply because practically never do the premises of such an explanation form a sufficient basis for the deduction of the conclusion.

The "documents" of historical geology are rocks, and the explicandum will consequently consist of a descriptive statement, or, more frequently, the conjunction of several descriptive statements, about the character and distribution of rocks. These statements have no historical dimension. They are about things that exist now and that can be verified more or less directly by observation. The statements may contain physical-chemical terms, mathematical terms, and specifically geological terms.

The tradition of a sound observational basis for historical geology is very old and in fact predates what everyone would agree was the beginning of modern geology. Werner, for example, contended that his system had a sound observational basis and, in the sense that it rested upon careful and detailed field observation and adequate descriptive statements, it did.

The great methodological problem of historical geology has never been the problem of framing adequate descriptive statements about what now exists, but the problem of how to meet the empirical condition of adequacy that the statements of the explicans of an explanation be "adequately supported by evidence based on data other than the observational data upon which the acceptance of the explicandum is based." Because statements about the past are inferred from statements about the present in terms of some general statement, it is clear that the major problem is how to place some "empirical" restriction upon the general statements that may be admitted to the explicans of a historical geological explanation. There is no test, however, to which a generalization can be put that will allow one to determine whether or not it has held throughout geologic time. The uniformitarian principle may be viewed as a methodological device or convention that limits the generalizations used in primary historical geological explanations to statements that meet the empirical requirements set for any valid generalization in science, that is, that they have been verified and have not been falsified here and now.

In the construction of an explanation of this sort observations are made that serve as the basis for the formulation of singular, descriptive statements about the character and distribution of rocks. These statements may be regarded as the explicandum. An attempt is then made to determine some singular antecedent statements that, together with certain pertinent generalizations admitted to the explanation with the assumption of uniformity, would constitute a satisfactory explicans with regard to this explicandum. The singular antecedent statements of the hypothetical explanation may come to be regarded as statements describing the geologic past. The procedure may be characterized in the words of Nagel quoted earlier: "We may note the occurrence of some event, and then seek to explain it by discovering some other event, which on the basis of an already established law, is assumed to be the condition for the occurrence of the given event." The operation is thus postdictive in character. In practically all cases it represents "inductive" rather than "deductive systematization," and perhaps could be called "probabilistic postdictive inference."[1]

A relatively uncomplicated example of this procedure may be found in Schoff (1951, p. 641): "The first orogeny affecting the Cedar Hills is inferred from the coarse conglomerates of the Indianola, which indicate large, active streams with high gradients and suggest that erosion had been accelerated by the folding or uplifting of mountains."

In this explanation the explicandum consists of statements about the character of the Indianola group. The singular antecedent statements of the explicans describe large active streams of steep gradient and high competence flowing from mountains. These two sets of statements could be connected only in terms of some implicit generalizations about the character of streams and the material that is transported by them and deposited from them. I am convinced that, in Schoff's explanation and in every other legitimate geologic explanation, the connection between specific events is made in terms of some generalization which can usually be produced should the framer of the explanation be pressed to "fill out the explanation sketch" or "produce role justifying grounds for the explanation." The probabilistic nature of the cited explanation is evident. The statement of antecedent conditions together with the implicit generalizations is clearly not a sufficient basis for the deduction of statements describing the precise character of the Indianola group or any part of it. Finally the explanation is implicitly, if not explicitly, "genetic" in the sense that a series of temporally prior and subsequent events are described or assumed. The minimum number of events involved in the explanation is: a conglomerate in the present, the deposition of a gravel at some past time, and at a still earlier time the beginning of orogenic movement.[2]

In making observations and in framing descriptive statements on the basis of these observations the geologist, like the historian, is selective. There are numerous factors that determine this selection; for example, personal preference, limitations of time, and arbitrarily imposed restrictions in space. The basis for selection, however, that is of overriding importance is imposed by the generalizations that are available, or to put it another way, by the nature and scope of contemporary physical-chemical and geological theory. Descriptive statements that are not covered by a geologic or physical generalization can serve no inferential purpose. Read (1952, p. 52) makes this observation:

> We use the exact methods of chemistry and physics to say *what* the earth's crust is made of, and we even attempt to guess what the deeper portions are like on the same kind of evidence. But the description of the stuff of the earth, fascinating though the description may be, is only incidentally a part of geology.

I agree with this point of view and would only add that if physical and chemical generalizations were not to be employed in the inferential procedure of reconstructing the earth's history, there would be no point in framing the descriptive statements in physical and chemical terms. And if the explicandum and the generalizations of the explicans are framed in physical-chemical terms, the singular antecedent statements of the explicans must be so framed.

Many of the generalizations employed in geologic inferences contain specific geologic terms, while others contain physical and chemical terms either exclusively or in addition to geologic terms. Physical and chemical generalizations may, in addition to being used directly in geological explanations—as, for example, in radioactive dating—serve as the basis for the explanation of generalizations containing geologic terms. Many geologists would go so far as to assert that all geologic generalizations are either in fact or in principle deductively derivable from the laws of physics and chemistry. There are, on the other hand, geologists who would deny this, but there is no school in geology comparable to the "holist" or "organismic" school in biology.

Many, if not most, of the generalizations that contain specifically geologic terms are probabilistic in character and contain such words as "normally" or "usually." For example: "If it [dust] settles on flat uplands where rainfall is only moderate and erosion is negligible, it may accumulate to form loess" (from Dunbar and Rodgers, 1957, p. 23). Others, however, are of universal form. For example: "If the supply of new sand is less than that moved over the crest, however, the windward slope will be degraded as the leeward slope grows, and thus the dune will migrate with the wind by transfer of sand from one slope to the other" (ibid., p. 20).

The probabilistic character of most geologic explanations results from the use of probabilistic generalizations. Even in the rare cases where strictly universal statements are employed in a geologic explanation, the explanation will almost always be probabilistic because a strict universal statement is regarded as holding only when a number of specific boundary conditions also hold, and almost never will the framer of a geologic explanation have knowledge of all the boundary conditions that limit the range of application of the universal statement which he has employed. I am in no position to judge whether or not Scriven's (1958) comment to the effect that virtually all scientific explanations are probabilistic is justified. This question need not concern us here, because, whatever the situation in science as a whole, geologic explanations are, practically without exception, *obviously* probabilistic in character.

Generalizations containing geologic terms have, for the most part, been arrived at "inductively" on the basis of a large number of observations of extant geologic phenomena and processes.[3] Geologists may also formulate general statements about historical succession. The significance of this type of generalization will be treated in a later section. In any case generalization about the present must be the prior operation.

### INDEPENDENT SUPPORT FOR THE ANTECEDENT STATEMENTS OF A HYPOTHETICAL HISTORICAL EXPLANATION

By convention the uniformitarian principle places an empirical restriction upon the general statements admitted to a hypothetical historical explanation. The empirical condition of adequacy will not technically be met, however, until the singular antecedent statements are supported by evidence that is independent of the evidence cited to support the explicandum. Confidence in a particular case of primary historical inference increases as this independent support for the antecedent statements is provided.

Independent support for the antecedent statements of a hypothetical explanation is provided through the construction of another explanation. Ideally this second explanation would contain an explicandum having no statements in common with the hypothetical explanation for whose antecedent statements support is being sought, but having identical antecedent statements, or at least pertinent antecedent statements in common, with this explanation.

Let me cite an explanation in which it is fairly clear that independent support for the antecedent conditions has not been provided.

Jenks and Goldich (1956, p. 169) state:

> The description of the sillar and of the major andesitic eruptive centers and the topographic relationship between the two features clearly indicate that the tuff flows originated in the same general centers as the andesite volcanoes. No specific rhyolitic eruptive centers have been discovered, and rhyolite lava flows are rare. Apparently the sources of the rhyolitic eruptives have been obscured by thick accumulations of andesite.

In this example the evidence presented to support the singular antecedent statements consists of a description of the phenomena to be explained together with certain implicit general statements. Until independent evidence is provided to support the antecedents the explanations clearly fail to meet the empirical conditions of adequacy and consequently must be regarded as technically hypothetical. It is hardly necessary to point out, I think, that no criticism of the authors is intended. It is perfectly obvious that they attempted to provide independent evidence for the antecedents and failed only because of particular sets of circumstances completely beyond their control. Many explanations in geology remain hypothetical for this reason. Jenks and Goldich in the phrase "clearly indicate" reveal a rather high degree of confidence in their explanation. It is not at all uncommon for geologists to have a high degree of confidence in a hypothetical situation. In some cases the degree of confidence in a hypothetical explanation may be so high that no effort is made to find independent support for the antecedent statements. In these cases the requirement of independent support may be considered excessive.[4]

Frequently it is possible to infer from a particular group of descriptive statements and certain selected general statements a functional relationship between two or more necessary antecedent variables, but not possible to infer particular values for each of the variables. In such cases independent evidence must be found that will allow the assignment of a particular value to one or more of the variables before a value can be assigned to the related variables. This situation often obtains in cases where physical and chemical laws, in which the geologist may have a high degree of confidence, are employed in historical inferences. Ingerson (1953, p. 304) considers this problem in the following passage:

> For many geological interpretations based on the distribution of isotopes, information is needed which is not immediately or directly available. In the laboratory it is comparatively simple to measure the factors involved —isotopic composition of original materials and products, temperature, rate of diffusion, distances, time, etc.—and to relate them to one another as desired for the specific purpose. In nature, however, it may not be possible to determine any factors except the present isotopic composition of the material, which may be the result of a reaction or process that took place half a billion or more years ago. The interpretation, then, becomes a game that involves evaluating by independent means the other factors essential to the solution of the problem—extrapolation, deduction, measurement of other properties of the material or associated substances, or just making assumptions, the accuracy of which varies tremendously according to the circumstances of the problem.

Turning to a specific problem, Ingerson goes on to observe (p. 355):

> Relations similar to those postulated for granitized areas would be expected for veins formed by percolating hydrothermal solutions. If the assumptions are correct, then determinations of isotopic composition along the strike and down dip of the vein should make possible determination of the direction from which the solutions came. This assumes that temperature of formation can be estimated independently from liquid inclusions, or by some other means.

In many geologic explanations it is clear that the antecedent statements have been strongly supported by evidential statements independent of the evidential statements upon which the explicandum is based. For example a detailed analysis of the joint systems of limestone, dolomite, and granite boulders contained in drumlins in Saskatchewan led Kupsch (1955) to conclude that the joints were produced by a simple compressive force acting on the wall of the drumlin and that such a force resulted from ice moving against the front of the drumlin. Whatever the adequacy of this explanation, it certainly does not fail to meet the condition that the antecedents be highly confirmed by evidence other than that upon which the acceptance of the explicandum is based. There is abundant evidence having nothing to do with jointed boulders that ice moved in Saskatchewan at a particular time and that it moved against the front of a drumlin.

In complex geologic explanations it is not always easy to decide whether independent evidence to support the antecedent statements has been provided or not. It is not uncommon that several or many detailed features of what is considered to be a geologic "unit," for example, of a formation, are offered as evidence to support a single, broad conclusion about antecedents. Fuller (1955), for example, lists ten features of the Sharon Conglomerate of Ohio that indicate a northern source for this unit. Although a single, defined lithologic unit constitutes the general body of evidence upon which the inference is based, each of the ten features does in itself, Fuller implicitly suggests, allow the construction of a hypothetical explanation with a northern source as a necessary antecedent condition. Each of the ten explanations provides evidence "other than" that offered by each of the other explanations for the support of the antecedents in that each explanation is formulated in terms of independent explicanda. Evidence containing no statements about the Sharon Conglomerate might be regarded as more reliable confirmation of such a source than a description of a separate feature of the formation itself, but technically one fulfills the empirical condition of adequacy as well as the other. The only unit that it is meaningful to recognize in such a case is the unit of evidence upon which each of the separate inferences is based. Whatever the utility of the concept "formation" in other connections, it has no significance here.

In the examples cited, evidence for the support of hypothetical antecedents was provided by two or more explanations formulated in terms of independent bodies of evidential statements about extant phenomena. It is sometimes possible to approach the problem of independent support for the antecedent statements in a somewhat different way by taking the hypothetical antecedents together with some selected general statements and from them to *predict* some previously unobserved, or at least unrecorded, present consequence. It is only in cases of this sort that anything like prediction plays a significant role in geology. As an example of this role of prediction I cite Swineford (1949) who concluded, in effect, that a necessary antecedent for the explanation of the volume, character, and distribution of the Pleistocene Pearlette ash was a large, explosive volcano that emitted acid glass during a particular interval of Pleistocene time and lay somewhere to the southwest of the area in which the ash had accumulated. She concluded from this that she might expect to find in a particular area evidence for the existence of this volcano that was completely independent of the evidence for the ash. In the Valle Grande caldera and its associated volcanic rocks Swineford suggests that she may have found such evidence.

The predictive character of this type of operation has become less obvious as geologic knowledge has increased. It could be argued, for instance, that Swineford did not predict that she would find independent evidence for the existence of the volcano but rather, since the existence of the caldera had been known for a long time, that she chose this particular body of evidence from a number of alternatives. A survey of earlier geologic literature reveals that the predictive aspect was more evident early in the history of modern geology than it is today. For example, in Hutton's *Observation on Granite* (1794, p. 77) we find the following comment:[5]

> Granite also was considered there [*Theory of the Earth*, 1788] as a body which had been certainly consolidated by heat; and which had, at least in some parts, been in the state of perfect fusion, and certain specimens were produced, from which I drew an argument in support of this conclusion.

In 1788 Hutton's views on the origin of granite had resulted from a hypothetical explanation framed on the basis of the character of individual specimens of granite and in terms of some generalizations about the process of crystallization.

Hutton goes on to say (p.78): "But my object was to know if the granite that is found in masses has been made to flow in the bowels of the earth, in like manner as those great bodies of our whinstone and porphyry, which may be considered as subterranean lavas."

Hutton thus sought independent evidence of the fluid origin of granite, that is, independent of the structure of isolated pieces of granite. He made a "prediction" as to the nature of this independent evidence in the statement (p. 78): "Now, the evidence of this must be found in the broken, separated and distorted parts of those regularly formed bodies, the natural history of which we so far know." The story of the dramatic verification of this prediction in Glen Tilt is well known and may be summarized in Hutton's words (p. 79): "I here had every satisfaction that it was possible to desire, having found the most perfect evidence, that granite had been made to break the Alpine strata, and invade that country in fluid state."

The choice among alternative hypothetical explanations, none of which violates geological or physical theory, will ideally be made on the basis of the independent evidence available to support the hypothetical antecedent statements of each explanation. The eighteenth-century explanation of the erratic boulders of northern Europe in terms of rafting by icebergs was rejected in favor of a glacial explanation because independent evidence to support the necessary antecedent statements was not forthcoming. In the case of the glacial explanation such evidence was eventually presented.

Controversy may rage for years among geologists because no decisive independent evidence can be presented to support the antecedent statements of any one of a number of alternative hypothetical explanations of a particular phenomenon. The controversy about the origin of the erratic boulders in the Johns Valley Formation of Oklahoma is an example of this kind of situation.

Choices among alternative explanations of geologic phenomena are not always made on the basis of independent evidence available for the support of antecedent statements, but are often made on the basis of available theory. For example, the hypothesis of continental drift has been rejected by many primarily for the reason that they believe the phenomenon is inconsistent with contemporary physical theory.

## THE EXPLANATION OF PAST EVENTS

Up to this point the discussion of explanation in geology has been restricted to cases in which the explicandum consists of statements about events and structures that can be observed in the present. In such cases the problem of empirical support for the statements of the explicandum is not a serious one. I have emphasized this particular application of explanatory inference because it must precede all others. It is the means by which we are justified in making statements about the past.

I wish to consider here the case of an explanation in which the explicandum is a statement about a past event that is related by means of appropriate general statements to another statement about a past event. The two statements so related are themselves the end products of two different primary historical inferences. Confidence in this kind of explanation may be less than in an explanation in which the explicandum is more or less directly testable by observation. The degree of confidence will depend upon the empirical support available for each of the events involved and upon the degree of confidence in the general statements used to connect the events. The general statements employed in explanations of past events are for the most part the same generalizations that are employed in primary historical inferences.

The explanation of past events in geology is important because it serves to connect events inferred on the basis of independent bodies of evidence into a logically integrated historical chronicle. The logical connection of some events is of course automatically provided in the process by which the events were inferred. It is consequently impossible to speak of a historical chronicle without any causal connection between events, because some causal connection is implicit in the process of primary inference.

The failure to state explicitly whatever generalizations are used to connect two or more events inferred on the basis of independent evidence is much more serious than the failure to state the generalizations employed in the explanation of some extant structure. In the latter case, as I stated earlier, I believe we must assume that some general statement has been employed to make the inference possible. In the former case, unless the generalizations are explicitly stated we are unable to conclude whether it is the author's intention simply to assert that one event followed another, or to claim that the two events are causally related.

"Geologic histories" usually represent, in part, at least, an attempt to connect independently inferred events into a system of historical or genetic explanations. This intention is clear in the summary of a paper by Shelton (1955, p. 88) on the Glendora volcanic rocks of the Los Angeles Basin, in which he states:

The basin began to subside in mid-Miocene time and reached its climax of depth and localization in the upper Miocene and Pliocene.
About the time of the Miocene deepening, and perhaps genetically related to it, volcanism began on the east side of the Basin, though farther west eruptions may have started earlier in the Miocene.

The suggestion that the two events are "perhaps genetically related" would presumably not have been made unless some generalization existed that related basinal subsidence and nearby volcanic activity.

In the following passage from a paper by Lowell and Klepper (1953, p. 242) on the Beaverhead formation of Montana, the intention to connect events into a genetic or causal sequence is again clear:

The Mesozoic and older formations were folded into generally north-trending anticlines and synclines during the Laramide orogeny. Erosion was rapid on the upper parts of rising folds. The detritus was deposited in near-by newly initiated intermontane basins as a thick sequence of dominant conglomerate and subordinate sandstone, siltstone, and limestone.

There are no general statements in this passage, yet no geologist reading it would doubt that it was the intention of the authors to *explain* the accumulation of certain kinds of sediments in terms of the formation of anticlines, and that furthermore perfectly familiar geologic generalizations could be cited which would make the explanatory connection logically legitimate.

In the following quotation from a paper on the geology of the Eastern Venezuela Basin by Hedberg (1950, p. 1211), it is not clear whether the author means to suggest that there is a causal or genetic connection between the several geologic events mentioned, each of which was presumably independently inferred, or not:

The latest Eocene or early Oligocene was a time of pronounced crustal movements over much of Venezuela. In western Venezuela the Venezuelan Andes were uplifted, separating the Maracaibo basin from the Apure basin; in Eastern Venezuela, similarly, folding and uplift appear to have taken place in the Parian borderland. At the same time there was a pronounced subsidence of the Eastern Venezuela Geosyncline with the shifting of its axis southward to the latitude of the present Serranía del Interior in northeast Venezuela and somewhat farther south in north-central Venezuela.

### *Historical Laws and the Explanation of Complex Events*

We may suspect that, because of their temporal and spatial association, two or more independently inferred past events are causally related. There may be no known generalizations or laws, however, by which the two events may be logically connected to form an explanation. As Nagel has put it (1961, p. 32), "We may note the occurrence of two or more events, suspect them to be significantly related, and attempt to discover the laws which formulate the specific modes of dependence between events of that character." If the spatial and temporal association that led to the hypothesis of causal connection between two events is found to be very frequent or constant between events of this kind, a "historical generalization" or "historical law" may be formulated that in essence simply consists of the assertion that events of this kind are in fact usually or invariably associated. Historical generalizations are frequently framed by geologists, but it is my observation that they are seldom if ever regarded as statements possessing explanatory power. In other words, they do not, in the opinion of geologists, "formulate the specific models of dependence between events of that character." It is quite evident that most generalizations of this kind were not proposed with inferential utility in mind but were rather meant to convey a complex body of historical information as economically as possible. Historical generalizations may be significant in that they serve as a guide for investigation directed toward the goal of formulating the specific modes of dependence in terms of more comprehensive geological generalizations and ultimately, perhaps, in terms of physical and chemical laws.

Let us consider an example from a paper by Keller (1953) on the clay minerals of the Morrison formation. He says (p. 102): "Because illite is the characteristic clay in a marine shale (Millot, 1949; Grim, Dietz and Bradley, 1949; Grim, 1951), its presence at first thought appears to be incompatible with a non-marine origin for the Morrison."

In other words if we regard the generalization about the association of marine environments and illite as having explanatory force, then it would clearly follow that the Morrison formation was of marine origin. Keller goes on to state: "However, a broader approach to the genesis of the clays emphasizes that indicator clays are formed as a response to a set of physico-chemical conditions rather than in an artificially classified environment, such as *marine*."

The association of a particular clay with a particular geologic environment expressed in a general statement, Keller contends, is not a sufficient formulation of their specific modes of dependence. He proceeds to analyze the specific modes of dependence in terms of physical-chemical theory.

One of the most widely accepted historical generalizations in modern geology is the assertion that geosynclines and orogenies are significantly related. Leet and Judson (1958, p. 342) have called it "one of the most significant generalizations in all geology." Despite almost universal agreement that geosynclines and orogenies are invariably associated in time and space and that the association is causally significant, no one has explicitly or implicitly suggested that the general statement expressing the relationship has explanatory force. Geologists clearly regard the formulation of this generalization not as an end but as a beginning, and are proceeding in an attempt to reduce it to comprehensive geologic generalizations and to physical and chemical laws.

Almost all of the generalizations of historical succession in geology concern complex events. Nagel (1961, p. 569) has this to say about the generalization of complex events in human history:

> It is rarely possible to account for a collective event which possesses an appreciable degree of complexity by regarding it as an instance of some recurring *type* of event, and then exhibiting its dependence upon antecedently existing conditions in the light of some (tacit or explicit) generalization about events of that type.

In geology, as in history, it is almost never possible to formulate a satisfactory generalization about the connection of complex events of one class with complex events of another class. Even though a number of events may be similar enough to be designated by the same general term, there are always demonstrable and significant differences among them, and for this reason each of the events may be regarded as unique. Some historians and philosophers have, in fact, denied the possibility of any meaningful generalization about the succession of historical events on the basis that each historical event is unique.

Without inquiring too deeply into the question of what constitutes a historical geological event, it may be said that only in the very simplest cases is it not possible to analyze such an event into a number of component events. The more complex the designated event, the easier it is to separate into components. Of events of the level of complexity of orogenies it is not only possible but apparently necessary to analyze them in this way. After such an analysis the problem of explaining a single complex event is removed, and the geologist may proceed in an attempt to offer explanations for each of the component events.

One needs only to consider again the attempt to explain the association of geosynclinal deposition and orogeny as an illustration of this procedure. The two events have been divided into a number of component events. A strenuous effort is being made to explain each of the component events and where possible to relate one component event to another.

It should be pointed out, I think, that to a large extent the recognition of component events precedes the recognition of the complex event. The complex historical event is actually a group of events that because of their frequent or invariable association have come to be regarded as a single unit. The separation into components is thus implicit in the recognition of the complex event. The recognition of the association may lead to the recognition of new components, or to put it another way, make possible inferences that lead to an increasingly detailed characterization of the event.

*Laws of Historical Succession*

It is not uncommon in historical investigations that a particular, pervasive, nonrepeatable trend may be recognized in historical succession. Occasionally a statement describing one of these trends may be called a "law." Popper (1957) and others have pointed out that the so-called "laws of historical succession" or "laws of evolution" are not "natural laws" in the usually accepted sense of this term because they consist, for the most part, of descriptions of sequences of unique events and nonrepeatable process. The process, Popper maintains, may proceed in accordance with certain laws. The description of the process is not a law, however, but a conjunction of singular historical statements. This confusing application of the term "law" does not have important consequences unless we proceed to frame predictions and explanations in terms of a "law" of historical succession.

"Laws" of historical succession in this sense are rare, if not nonexistent, in geology. Descriptions of trends, for example, of continental accretion, are not uncommon in geology, but I know of no instance of their having been called "laws," nor do I know of any case in which an explanation has been framed in terms of them. Geologists who have attacked the problem of continental accretion have proceeded in typical fashion in an attempt to explain it in terms of geological and physical generalizations.

## CONCLUSION

To me the similarity between historical explanation and geologic explanation is striking. In addition to the primary concern in both disciplines with the explanation of particular events, geologic and historical explanations are commonly alike in being probabilistic, genetic, and "sketchy," particularly as a result of a lack of the explicit statement of the generalizations employed.

The failure of geologic explanations to meet all of the rigorous conditions of adequacy set for a scientific deductive explanation is apparently what leads many geologists to remark upon the inexactitude and uncertain character of their conclusions when compared to the conclusions reached by other scientists. Read (1952, p. 58) reflects this attitude in the following quotation:

> My purpose is to indicate that there are geological arguments and methods of thought, which though based upon a combination of dimly perceived facts, partly controlled guesses, personal intuitions and all manner of nebulous factors, not excluding a species of low cunning, are yet in a geological matter decisive.

It is my conviction that to make the inferential procedures of our discipline explicit in light of the clear recognition of its historical character may lead to increased consistency, economy, and completeness in the system of geology.

## References Cited

AYER, A. J., 1958, The problem of knowledge: London, Macmillan & Co. Ltd., x + 258 p.

COHEN, M. R., 1947, The meaning of human history: La Salle, Ill., Open Court Publishing Co., ix + 304 p.

DEWEY, J., 1938, Logic, the theory of inquiry: New York, Henry Holt & Co., vii + 546 p.

DUNBAR, C. O., and RODGERS, J., 1957, Principles of stratigraphy: New York, John Wiley & Sons, xii + 356 p.

ELLISON, S. P., 1960, Thinking patterns for geologists: Am. Assoc. Petroleum Geologists Bull., v. 44, p. 980-983.

FRANKEL, C., 1957, Explanation and interpretation in history: Philosophy of Science, v. 24, p. 137-155.

FULLER, J. O., 1955, Source of Sharon Conglomerate of northeastern Ohio: Geol. Soc. America Bull., v. 66, p. 159-176.

GALLIE, W. B., 1955, Explanation in history and the genetic sciences: Mind, v. 64, p. 160-180.

GARDINER, P., 1952, The nature of historical explanation: London, Oxford University Press, xii + 142 p.

HEDBERG, H. D., 1950, Geology of the Eastern Venezuela Basin (Anzoategui-Monagas-Sucre-Eastern Guarco portion): Geol. Soc. America Bull., v. 61, p. 1173-1215.

HEMPEL, C. G., 1942, The function of general laws in history: Jour. Philosophy, v. 39, p. 35-48.

———, 1958, The theoretician's dilemma, a study in the logic of theory construction, *in* FEIGL, H., *et al.* (eds.), Minnesota studies in the philosophy of science, v. 2: Minneapolis, University of Minnesota Press, xv + 553 p.

——— and OPPENHEIM, P., 1948, Studies in the logic of explanation: Philosophy of Science, v. 15, p. 135-175.

HUTTON, J., 1788, Theory of the earth: Royal Soc. Edinburgh Trans., v. 1, p. 209-304.

———, 1794, Observations on granite: *ibid.*, v. 3, p. 77-85.

INGERSON, E., 1953, Nonradiogenic isotopes in geology: a review: Geol. Soc. America Bull., v. 64, p. 301-374.

JENKS, W. F., and GOLDICH, S. S., 1956, Rhyolitic tuff flows in southern Peru: Jour. Geology, v. 64, p. 156-172.

KELLER, W. D., 1953, Clay minerals in the type section of the Morrison Formation: Jour. Sed. Petrology, v. 23, p. 93-105.

KUPSCH, W. O., 1955, Drumlins with jointed boulders near Dollard, Saskatchewan: Geol. Soc. America Bull., v. 66, p. 327-338.

LEET, L. D., and JUDSON, S., 1958, Physical geology: 2d ed.; Englewood Cliffs, N.J., Prentice-Hall, Inc., ix + 502 p.

LOWELL, W. R., and KLEPPER, M. R., 1953, Beaverhead formation, a Laramide deposit in Beaverhead County, Montana: Geol. Soc. America Bull., v. 64, p. 235-244.

MELTON, F. A., 1947, Proposed monograph of historical geology and the geological education controversy: Geol. Soc. America Interim Proc., pt. 1, p. 49-58.

NAGEL, E., 1952, The logic of historical analysis: Sci. Monthly, v. 74, p. 162-169.

———, 1961, The structure of science: New York, Harcourt, Brace & World, Inc., xiii + 618 p.

POPPER, K. R., 1957, The poverty of historicism: Boston, Beacon Press, xiv + 166 p.

———, 1959, The logic of scientific discovery: New York, Basic Books, 480 p.

READ, H. H., 1952, The geologist as historian, *in* Scientific objectives: London, Butterworths Scientific Publications, vii + 113 p.

RENIER, G. J., 1950, History—its purpose and method: Boston, Beacon Press, 272 p.

SCHOFF, S. L., 1951, Geology of the Cedar Hills, Utah: Geol. Soc. America Bull., v. 62, p. 619-646.

SCRIVEN, M., 1958, Definitions, explanations, and theories, *in* FEIGL, H., *et al.* (eds.), Minnesota Studies in the Philosophy of Science, v. 2: Minneapolis, University of Minnesota Press, xv + 553 p.

———, 1959, Truisms as the grounds for historical explanations, *in* GARDINER, PATRICK (ed.), Theories of history: Glencoe, Ill., Free Press, ix + 549 p.

SHELTON, J. S., 1955, Glendora volcanic rocks, Los Angeles Basin, California: Geol. Soc. America Bull., v. 66, p. 45-90.

SWINEFORD, A., 1949, Source area of Great Plains Pleistocene volcanic ash: Jour. Geology, v. 57, p. 307-311.

WALSH, W. K., 1951, Introductions to the philosophy of history: London, Hutchinson & Co., Ltd., 174 p.

CHAPTER TWO

# Certainty and Uncertainty in Geology

## INTRODUCTION

There has been a good deal of discussion during the past century about the provisional character of geologic knowledge (see, for example, Chamberlin, 1904; and Bradley, 1963). This discussion has included comparisons between uncertainty in geology and uncertainty in physics which have, inevitably it seems, led to the question of determinism and indeterminism in nature (see, for example, Leopold and Langbein, 1963; Krauskopf, 1968; and Mann, 1970).

Although treatments of uncertainty in geology have purported to deal, at least in part, with methodological issues, the metaphysical question of determinism has both detracted from and been confused with purely methodological issues. In this discussion I propose to consider the character of hypotheses in which geological uncertainty is expressed and the role of these hypotheses in the explanation, prediction, and retrodiction of events. I shall consider determinism only in the hope of demonstrating its irrelevance to methodological problems in geology. Before embarking upon a discussion of uncertainty, let us briefly consider certainty, insofar as it is supposed to exist, in science.

## DEDUCTIVE EXPLANATION AND SCIENTIFIC LAWS

### *A*

For most writers the notion of "cause" lies at the heart of any consideration of certainty in science. What can be said of this concept, bearing in mind my intention to confine attention to methodological issues? Popper has said (1957, p. 122), "I suggest that to give causal explanation of a certain *specific event* means deducing a statement describing this event from two kinds of premises: from some *universal laws*, and from some singular or specific statements which we may call the *specific initial conditions*."

According to Popper, there must be a very special kind of connection between the premises and the conclusion of a causal explanation. It must be deductive. In a deductive argument, the conclusion follows *necessarily* from the premises. The premises and the conclusion consist, as Popper makes clear, of statements. Lest the notion of "necessity" be misunderstood in this context, it should be pointed out that it is imparted by the correct application of the rules of deductive logic, rules that are concerned with the relationship among statements. We must be prepared to justify every step in a deductive argument by citing such a rule. We may feel that there is a very strong, even a necessary connection between two *events*. If we are to justify such a connection logically, it must be on the basis of a rule connecting *statements that describe these events*. There is no rule of logic that permits us to make a necessary connection between a statement describing one particular event and a statement describing another. If an argument is to yield a conclusion in the form of an unconditional singular statement, its premises must consist of at least one universal statement *and* at least one singular descriptive statement. Hempel and Oppenheim (1948) point out that the requirement of a deductive link between the premises and the conclusion of an explanation amounts to a requirement that an explanation could have served as a prediction if the premises had been known before the event to be explained occurred.

The concept of "law" lies at the heart of deductive explanation and therefore at the heart of the certainty of our knowledge about specific events. Bradley, in his discussion of geologic laws (1963), accepts Pearson's characterization of a scientific law, as, "a *resumé* in mental shorthand, which replaces for us a lengthy description of the sequences among our sense-impressions" (Pearson, 1900, p. 86-87). But if scientific laws were no more than summaries of the cases we have examined, there would be no *problem of scientific laws*. A problem arises because scientists are prepared in the name of laws to make assertions about unexamined cases. A law could not be used to justify a prediction if it were no more than a resumé of all the relevant cases that had ever been encountered. The critical feature of a prediction is that it has something to say about a case that has not yet been encountered. To fulfill the function required of them, scientific laws must be of "essentially generalized form" (see Hempel, 1965, p. 340), which is to say they must not be equivalent to a finite conjunction of singular descriptive statements. The following statement from Robinson (1972, p. 1693) is of universal form because it refers to *all* the members of a class: "All of the volcanic rocks of the Silver Peak center belong to a potassic suite ranging from trachybasalt to rhyolite." It cannot be used to justify an assertion about an unexamined case. All the cases within its scope have already been examined. No one will disagree with Nagel who says (1961, p. 63), "But if a statement asserts in effect no more than what is asserted by the evidence for it, we are being slightly absurd when we employ the statement for explaining or predicting anything included in this evidence." Robinson's statement is truly "a resumé in mental shorthand." It could be eliminated from Robinson's description of the volcanic rocks of the Silver Peak region in favor of a finite conjunction of singular sentences without changing the meaning of that description. Although universal in form it is not, nor is it intended to be, a law.

If a statement is to function as a law, it is critical that it be of essentially generalized form. But this has not been regarded as a sufficient condition for labeling a statement "scientific law." The term "law" has been reserved for statements of very broad scope, and it is usually a condition that statements that are accorded such status have no explicit or implicit reference to particular regions of space or time, or to particular objects. For the purposes of this discussion a scientific law may be defined as a statement of universal and essentially generalized form that makes no essential reference to individual objects, particular regions of space, or particular intervals of time. It shall further be required of a law that it be supported by empirical evidence. This will be taken to mean that within the context of an agreed-upon background of general knowledge some instances support the statement and no instances falsify it.

Essentially generalized universal statements are not as common in science as some accounts would lead us to believe. They are found in the most rigorously formulated physical theories and seldom anywhere else. The laws of classical mechanics are usually regarded as statements of universal form. We may interpret the *inertial law* as a statement to the effect that *whenever* and *wherever* a body is free of impressed force it will remain in its present state of motion within a designated spatial and temporal frame. As many authors have pointed out (see, for example, Nagel, 1961), there is something a little strange about this law because the answer to the question "Where and when is a body free of impressed force *in the sense required by the theory*?" is "Nowhere and at no time." In an important sense the universal propositions of classical mechanics do not describe concrete physical situations. The rigor and universality of these laws is assured by formulating them for the highly idealized case. In order to apply them to physical objects, it is necessary to introduce statements concerning initial and boundary conditions that cannot be derived from the laws themselves. The addition of supplementary assumptions concerning initial and boundary conditions does not *logically* entail uncertainty, but empirically it almost invariably does, simply because precise knowledge of these conditions is impossible to obtain.

*B*

Mann (1970) holds that the inability to predict specific events may stem from the fact that "nature is intrinsically random." According to Mann (p. 95), " 'Random' and 'randomness' are used in this paper in a statistical sense to describe any phenomenon which is unpredictable with any degree of certainty for a specific event" and "deterministic phenomena, on the contrary, are those in which outcomes of the individual events are predictable with complete certainty under any given set of circumstances if the required initial conditions are known." Mann does not give a final answer to that question "Is *nature* random?" although it is quite clear that he is inclined to think that it is. He expects that the question will be answered, for he says (p. 102), "Randomness has been suggested as the ultimate and most profound physical concept of nature. Time and sound scientific investigation will reveal if this is the case."

If nature were "intrinsically random" prediction would be impossible, but the continued failure to predict does not settle the issue of determinism. Of classical mechanics, Nagel says (1961, p. 285),

> Moreover, it is almost truistic to claim that classical mechanics is not a deterministic theory if the claim simply means that actual measurements confirm the predictions of the theory only approximately or only within certain statistically expressed limits. Any theory formulated, as is classical mechanics, in terms of magnitudes capable of mathematically continuous variation must, in the nature of the case, be statistical and not quite deterministic in this sense. For the numerical values of physical magnitudes (such as velocity) obtained by experimental measurement never form a mathematically continuous series; and any set of values so obtained will show some dispersion around values calculated from the theory. Nevertheless, a theory is labeled properly as a deterministic one if analysis of its internal structure shows that the theoretical state of a system at one instant logically determines a unique state of that system, for any other instant. In this sense, and with respect to the theoretically defined mechanical states of systems, mechanics is unquestionably a deterministic theory.

And he adds (p. 285 n), "In point of fact, it is on this basis that in current discussions theories are labeled as deterministic or indeterministic, and not on the basis of an examination of the experimental data which support them."

It is for this reason that when a predictor is introduced into a philosophical discussion of determinism it is not a human being but a "superhuman intelligence" (see Laplace, 1840). Human predictors cannot settle the issue of determinism because they are unable to predict physical events no matter what the world is *really* like.

*C*

The methodological force of claims about geologic uncertainty seems to be that deductive explanations and predictions of particular events are rare or absent in geological discourse. The requisites for deductive rigor are indeed lacking. An examination of the literature reveals that there are virtually no geological generalizations formulated in universal terms. Before taking some risky and unproductive leap to metaphysical assertions about the structure of the real world on the basis of this obvious fact let us attempt, in methodological terms, to understand why this is the case.

The uncertainty in geological knowledge arises out of the conviction that geological generalizations are immensely complicated instanciations of abstract, and often universal, physical laws. Geological generalizations *always* contain the assumption of boundary conditions. Geologists cannot predict what is going to occur because they have assumed boundary conditions beyond their ability to control them with certainty. They have done this because, on the assumption of physical theory, geologically significant configurations must be regarded as highly complex. This is true whether or not the "world" is determined. Physical laws, which are *not* formulated as universal statements, may impose uncertainty directly upon geology as in the case of inferences based upon the principles of radioactive disintegration.

## PROBABILITY STATEMENTS

*A*

It has been generally supposed that the geologic uncertainty alluded to above is expressed in probabilistic generalizations. Before considering probability inferences in geology let us consider what it is that a probability statement, any probability statement, is meant to convey. "The probability of throwing an ace with a single die is one sixth" might be taken in a *statistical* sense to mean that in a long series of trials an ace will be thrown with a frequency of approximately one time in six. On this interpretation the statement says nothing about a particular throw except insofar as it is a member of the class of events for which the specified frequently is said to hold. The statement could be used as the basis of a "prediction" that aces will continue to appear with a frequency of one sixth if the statement is supposed to be essentially generalized in the sense that the frequency holds for a series of unexamined instances. But if the statement is so interpreted, then the so-called prediction is simply an explicit expression of what is understood to be implicit in the statement itself. It is analogous to saying that from the expression "F = ma" it can be "predicted" that force will continue to have a value equal to the product of the values for mass and acceleration. There is nothing mistaken in such assertions. It can be said of them, however, that they do not get at history because they make no reference to particular events.

It is possible to interpret the statement "The probability of throwing an ace with a single die is one sixth" in such a way as not to lose hold of particular events. It can be taken to mean that our *knowledge* that in a long series of throws the frequency of aces is about one in six confers upon the hypothesis "An ace will be thrown on this occasion" a certain "probability." "Probability" cannot here be referring to frequency in the long run. It does not characterize a relationship among events, as does a statistical statement, but rather a relationship among statements. It is therefore *logical* or *inductive* probability.

A probable inference will include as one of its *premises* a frequency hypothesis, an assertion that if conditions of a specified kind are realized then conditions of another specified kind will be realized with a certain frequency in the long run. In arguments that invoke such hypotheses the conclusion cannot follow deductively from the premises, or to put it another way, the truth of the premises does not guarantee the truth of the conclusion. A probable inference in this strict sense is not simply an inference that yields an uncertain conclusion, but one that permits a certain degree of probability or "rational credibility" to be assigned to the conclusion. It is a remarkable fact that despite all the claims about the probabilistic nature of geological knowledge, explanatory probable inferences and the hypotheses that might support them are rare in geology. This obvious fact is often overlooked, apparently because mathematical statistics are so often applied to geological problems.

Consider the work of Potter and Blakely (1968), who examine lithologic sequences as dependent Markov processes. They say (p. 154), "Bedding sequences and stratigraphic sections can be profitably considered from the standpoint of their lithologic transitions, that is, the probability of any particular unit being followed by the other units in the sedimentation system." And they conclude (p. 164),

> These three examples strongly suggest that dependence in lithologic transitions may well be the rule rather than the exception. Geological support for this statistical conclusion is based on the idea that geologic sections are commonly generated by migrating environments which maintain fairly fixed

lateral positions relative to one another. Such migration is deemed to cause the vertical dependence of lithologic sequences.

The authors do not provide a statistical hypothesis to be invoked in explanations and predictions of stratigraphic sequences, partly, we may guess, because they have examined so few instances. But if analyses were performed on a large number of sequences, would they then be prepared to provide us with a probability hypothesis that might serve, for example, in support of a hypothesis that a sandstone would be followed in sequence by a shale? I don't think that they would. The analysis is not undertaken to provide hypotheses to be invoked in probabilistic inferences, but rather it is undertaken to provide descriptions of phenomena to be explained in terms of higher level hypotheses concerning the accumulation of sediments in migrating environments. *Ultimate* theoretical understanding lies in physical theory which may or may not be probabilistically formulated.

Mann's (1970) reaction to the work of Potter and Blakely is puzzling. He remarks (p. 98),

> Numerous recent investigations into duplication of sedimentological sequences by Markovian processes (Potter and Blakely, 1968; Carr and others, 1966; Krumbein, 1968; Vistulius and Faas, 1965) indicate that even the apparent regularity exhibited in stratification, including cyclotherms, may be randomly induced. Results of these workers certainly should not be interpreted as proving the presence of a stochastic process. However, we are alerted that cyclic sedimentation which so long has been construed as an obviously deterministic phenomenon—in fact, may not be.

Potter and Blakely have suggested that their analysis reveals a regularity in sequences that are overtly irregular, which is to say they show evidence of at least one step dependence or "memory" in the stratigraphic sequences they analyze. No matter what analysis revealed, it is difficult to see that it would have anything to do with whether or not the depositional events in question were "deterministic." If a sequence of depositional events turns out upon analysis to be a Markov chain, then the hypothesis that an event is dependent to some degree upon the next event in the sequence is supported. If analysis reveals the sequence not to be a Markov chain, then the hypothesis of dependence of the events in question is falsified. To show that an event is not dependent upon some specific event that preceded it demonstrates neither that it is dependent upon no event whatever nor that it was undetermined.

This point is illustrated in the work of Knopoff on the occurrence of earthquakes. He says of his study of the possible correlation of earth tides and earthquakes (1964a, p. 1869), "The conclusion from these calculations is that the largest possible triggering mechanism in the earth, namely that of oscillatory tidal strains, has no detectable influence upon the times of occurrence of small earthquakes in Southern California." Mann (1970, p. 98) cites this paper in support of the totally unwarranted conclusion that "the distribution of seismic events and energy release in time, and presumably in space, are randomly distributed." The *warranted* conclusion is just the conclusion Knopoff draws: that the seismic events are not related to earth tides, not that they are related to nothing whatever. In a paper that immediately follows the one cited above Knopoff (1964b) finds, he thinks, a relationship involving another kind of event. He says (p. 1873),

> The conclusion resulting from this calculation is that earthquake events of magnitude 3 or greater or 4 or greater in Southern California do not fit a Poisson distribution. This suggests, in a negative way, that a causal connection may exist between two successive earthquake events in Southern California; i.e., that the time of occurrence of an earthquake event is not completely independent of the time of occurrence of its predecessor or predecessors.

In the face of this Mann (p. 98) admits that the degree of randomness is "not truly simple."

## NORMIC STATEMENTS

Geology contains as few statistical hypotheses as it does universal laws. Perhaps those who deny this base their conclusion on the assumption that geological generalizations that express uncertainty are loosely formulated statistical hypotheses; that the "usually" and "probably" and "tends to" of these statements are meant, in the course of further investigation, to be replaced by expressions of frequency. This may be true in some cases. In other cases it is clearly not. Neither in geology nor in everyday life are all expressions of uncertainty to be regarded as covert or badly formed statistical statements.

Suppose that a rifleman were to report that from a particular benchrest position a certain rifle *tends* to put a bullet within a 5 cm circle on a target located 100 m from the rifle. In response to a suggestion that the uncertainty conveyed by the expression "tends to" could be reduced in the course of further investigation, the rifleman might fire a series of bullets into the target, counting those that fell within the circle and those that did not. On the basis of this count he might report that rifle $r$ from benchrest position $p$ puts bullets in a 5 cm circle at 100 m with a frequency $f$. If the rifleman did respond in this way, it would be plausible to suppose that he had intended his original report to be taken as an imprecisely formulated frequency hypothesis.

Consider now the statement, "Wind-driven sand, like snow, tends to settle in the wind-shadow of topographic obstructions" (Dunbar and Rodgers, 1957, p. 19). A geologist would not respond to a suggestion that he undertake an investigation to reduce the uncertainty expressed by the "tends to" of this statement by counting instances of deposition of sand in wind-shadows as against instances of nondeposition of sand in wind-shadows. The reason that he would not is clear. Frequency hypotheses are considered significant only when initial and boundary conditions are controlled or known to some minimum degree. The acceptable minimum degree of control over boundary conditions is determined by the investigator's knowledge, according to some theory, of what the relevant boundary conditions are, and how well he can determine them in a given case. The "tends to" in the statement of the rifleman presumably refers to an uncertainty that remains after initial and boundary conditions have been controlled as much as they can be or as much as it is appropriate to do. In the statement of Dunbar and Rodgers the "tends to" presumably refers not only to an uncertainty that would remain after the boundary conditions had been controlled as much as possible, but also to an uncertainty that stems from the values of a large unspecified variety of boundary conditions. The scope of the statement is the virtually infinite number of combinations of variables within the limits defined by "wind," "sand," "topographic obstruction," and "wind-shadow." When all of its instances are considered, the statement seems to imply the frequency of deposition is greater than one half.

But a geologist reads much more into the statement than just this. He understands, among other things, that the "tends to" holds *only* over the entire scope of the statement and may not apply to some designated part of the scope. If a geologist were asked, "If fine-grained sand driven by a wind of velocity 45 m per second and concentrated one grain to the cubic meter were to encounter a hemispheric topographic obstruction 2 cm in diameter, would some of it settle in the wind-shadow of that obstruction?" the answer to the question would be "No, never." Yet, on the other hand, if the geologist were asked, "If medium-grained sand driven by a wind of velocity 15 m per second and concentrated ten grains to the cubic cm were to encounter a cuboid topographic obstruction one meter in height, would some of it settle in the wind-shadow of that obstruction?" the answer to the question would probably be, "Yes, always." The geologist takes the statement to imply that under some combinations of initial and boundary conditions sand *never* settles in the wind-shadow of a topographic obstruction, and that under other combinations of initial and boundary conditions some of it *always* does. The statement does not explicitly delimit the two classes. The geologist is able to do this to some extent on the basis of his previous knowledge. This does not remove all uncertainty. In the case under consideration, and in most other geological generalizations, an area of uncertainty remains in the form of a class of combinations of initial and boundary to which "tends to" will still apply. But the degree of uncertainty about the deposition of wind-driven sand is much lower than the statement taken by itself would indicate. The statement as it stands may not enter directly into explanatory inferences but may serve rather as a summary and reminder of other statements that do.

Scriven has suggested that the *normic statement* expresses a kind of uncertainty that is not probabilistic (for an earlier discussion of normic statements in geology see Kitts, 1963). Scriven says (1959a, p. 466),

> The statistical statement is less informative than the normic statement in a very important way, although an exact statistical statement may be informative in a way a normic statement is not. The statistical statement does not say anything about the things to which it refers except that some do and some do not fall into a certain category. The normic statement says that *everything* falls into a certain category *except* those of which *certain special conditions apply*. And although the normic statement itself does not explicitly list what count as exceptional conditions, it employs a vocabulary which reminds us of our knowledge of this, our trained judgment of exceptions.

And he adds (p. 466),

> Now if the exceptions were few in number and readily described, one could convert a normic statement into an exact generalization by listing them. Normic statements are useful where the system of exceptions, although perfectly comprehensible in the sense that one can learn how to judge their relevance, is exceedingly complex. We see immediately the analogy with—in fact, the normic character of—physical laws. The physicist's *training* makes him aware of the system of exceptions and inaccuracies, which, if simpler, could be put explicitly in the statement of scope.

The normic character of some geological generalizations is obvious. For example, Weller states (1960, p. 152), "Carbonaceous material is a residue that remains after various more or less mobile compounds, produced during the decomposition of organic matter in an oxygen-deficient environment have moved away. Methane gas, for example, evolves and escapes under ordinary conditions." In this example the exceptional circumstances under which the generalization will not hold are not specified.

The following statement specifies at least some of the exceptional circumstances which, if they should apply, would invalidate the generalization. "Ordinarily the more resistant minerals, above all quartz, are concentrated on the beach, but there are many exceptions where non-resistant minerals are common or even dominant, as for instance where waves are relatively weak or have been at work only a short time or where quartz is uncommon or even lacking in the source area" (Dunbar and Rodgers, 1957, p. 64).

Normic statements are, I believe, very common in geology, although many of them may appear at first glance to be probabilistic. Terms such as "tends to" must not be taken at their face value. Even statements that express no explicit uncertainty may often best be interpreted as implicitly normic. Consider the statement, "Wind blowing across an outwash body sets sand grains into saltatory motion" (Flint, 1957, p. 185). There are clearly circumstances under which this statement does not hold, and in fact Bagnold (1941, p. 91) lists some of them.

The normic statement asserts that under "normal" or "standard" conditions an event of a particular kind will always be followed by an event of another particular kind. Where these conditions are not specified there is presumably some understanding as to what they are. As Hempel (1965, p. 348) points out, however, "But to the extent that those conditions remain indeterminate, a general statement of causal connection amounts at best to the vague claim that *there are* certain further unspecified background conditions whose explicit mention in the given statement would yield a truly general law connecting the 'cause' and the 'effect' in question." In many cases the conditions under which a geological generalization would hold are not specified because they cannot be. In other cases, however, they could be. The conditions under which wind blowing across an outwash body would *always* set sand into saltatory motion could be specified. The specification would be very complex, but it might permit the formulation of a universal statement. But such a statement would be of no use whatever under circumstances where the relevant initial and boundary conditions could not be determined, just as the gravitational equation is of no use under circumstances where mass, distance, and force cannot be determined. It is very rare indeed that values for all the relevant initial and boundary conditions can be obtained within the context of a geologic inference, and for this reason generalizations are, I believe, intentionally left in a loosely formulated state. It is in this state that they find their widest applicability.

This does not preclude the possibility that relevant specific conditions can be taken into account under favorable circumstances. It is clear, however, that the normic generalization with its implicit reference to initial and boundary conditions identified by theories which are, by universal agreement, applicable to geologic events, often gets at the sort of uncertainty encountered in geology far better than does the frequency hypothesis.

## RETRODICTION

The problem of uncertainty in geology cannot be solved on the basis of an examination of geologic generalizations. The formulation of generalizations is not the goal of geology. The goal of geology is the derivation and testing of singular descriptive statements about the past. To get at uncertainty in geology, we must consider the use to which geological generalizations are put in everyday practice. Retrodiction, not prediction, is the most characteristic geologic inference.

Discussions of explanation in history often begin with past events. But past events must be inferred. It is important to be reminded from time to time that the propositions with which historical inferences begin are not themselves historical but are instead assertions about the present. I have discussed the nature of historical inference in geology elsewhere (see above, pp. 12-20; below, pp. 77-82; and Kitts, 1963).

Our willingness to accept a geologist's account of the past results in part from the fact that he can often present more than a single inference to support his contention that certain events have occurred. He thus avoids circularity by meeting the condition that in adequate explanatory inferences the premises be supported by empirical evidence independent of the evidence presented in support of the event to be explained. But this is not always a critical factor. Our confidence in an inference is sometimes so high that the lack of independent support for its conclusions is of little consequence. Scriven's paresis case may illuminate this fact. He says (1959b, p. 480),

> Here, we can explain but not predict, whenever we have a proposition of the form "The only cause of X is A" (I)—for example "The only cause of paresis is syphilis." Notice that this is perfectly compatible with the statement that A is often not followed by X—in fact, very few syphilitics develop paresis. Hence, when A is observed, we can predict that X is *more* likely to occur than without A, but still extremely unlikely. So we must, on the evidence, still predict that it will *not* occur. But if it does, we can appeal to (I) to provide and guarantee our explanation.

And Hempel replies (1965, pp. 369-90),

Thus we have here a presumptive explanation which indeed is not adequate as a potential prediction. But precisely because paresis is such a rare sequel of syphilis, prior syphilitic infection surely cannot by itself provide an adequate explanation for it. A condition that is nomically necessary for the occurrence of an event does not, in general, explain it; or else we would be able to explain a man's winning the first prize in the Irish sweepstakes by pointing out that he had previously bought a ticket, and that only a person who owns a ticket can win the first prize.

Scriven's paresis case is not an example of an explanation in *Hempel's definition of explanation*, because it provides only a necessary and not the sufficient conditions for the occurrence of paresis. But the failure to provide a *Hempel explanation* does not prevent the valid inference of this necessary antecedent condition. Given the generalization about the connection between syphilis and paresis, and given someone with paresis, it follows necessarily that the person has a history of syphilis. The significant factor here is not our ability to provide independent empirical support for the antecedent conditions, but our confidence in the generalization. "Oxidation of ore deposits may occur without attendant sulphide enrichment, but enrichment cannot take place without accompanying oxidation" (Bateman, 1950, p. 274) is a statement of the same form as the generalization in Scriven's paresis example, and like it will support neither a Hempel explanation nor a prediction but will support a retrodiction.

Let us take another generalization from geology which I take to be normic. "If these fissures remain open while the surface is being buried and persist after the surrounding sediment has hardened to a rock, they constitute a trustworthy indicator of the top surface of the layer they penetrate" (Shrock, 1948, p. 189). We can easily imagine observing some newly formed mudcracks with this knowledge in mind and thinking, "If this surface is buried, and if the fissures remain open, and if they persist while the surrounding sediment hardens, and if the whole mass is not then removed by erosion, and if it is not altered beyond recognition by heat and pressure, and if the overburden is removed, then some geologist might encounter it and be able to recognize the top surface of the layer." Clearly this does not constitute a prediction. But let us suppose we find some structures that we recognize as fossil mudcracks. We may retrodict with a high degree of certainty that some mudcracks did form, and that they did remain open, and that they were buried, and that they did persist while the surrounding sediment hardened, and that the whole mass was not removed by erosion or altered beyond recognition by heat and pressure, and that finally the overburden had been removed to reveal the structures.

Events whose occurrence is covered by statistical hypotheses may sometimes be retrodicted although they could not have been predicted. Given, for example, the encounter of a *Drosophila* egg with a group of sperm we cannot predict whether the egg will be fertilized by a sperm carrying a Y chromosome or by a sperm carrying an X chromosome. We can only say that the probability of each is about equal. If, however, the fertilized egg develops into a male we can say that it was *certainly* fertilized by a Y chromosome bearing sperm, and if the egg develops into a female we can say that it was *certainly* fertilized by an X chromosome bearing sperm. Similarly the laws of physics do not permit us to predict which $U_{238}$ atoms *will* disintegrate, but they permit us to retrodict which ones *did* disintegrate. To take an example of geological interest, consider the following statement from Leopold and Langbein (1963, p. 190):

> Imagine a broad hill slope of uniform material and constant slope subjected to the same conditions of rainfall, an ideal case not realized in nature. Assume that the slope, material and precipitation were such that a large number of rills existed on the surface in the form of a shallow drainage net. Would it be supposed that rills comparable in size and position were absolutely identical? The postulate of indeterminacy would suggest that they would be very similar but not identical. A statistical variation would exist, with a small standard deviation to be sure, but the lack of identity would reflect the chance variation among various examples, even under uniform conditions.

It would not be possible to predict the path that a drop of water falling on the slope would take. I cannot imagine a geologist wanting to do so. There are a number of possibilities among which we could not choose with certainty. After the rill had developed, however, we might be able to determine the path that the water had in fact taken; an item of information that might be of considerable geological interest.

## RETRODICTIVE UNCERTAINTY

Hypotheses to the effect that the same kinds of initial conditions can result in different kinds of outcomes are the foundation of predictive uncertainty. After the occurrence of an event designated as a possible outcome by such a hypothesis, we may be able to determine directly or inferentially what the outcome was. This determination would not depend upon the frequency hypothesis that covers the occurrence of such events. If it were to be said of a *Drosophila* embryo that it *probably* resulted from an egg fertilized by a Y chromosome bearing sperm, the "probably" might not refer to the frequency with which eggs are fertilized with different kinds of sperm but rather to our confidence in our ability to determine the sex of an embryo at some early stage in its development.

To pursue this point, consider the following example. We bet on the outcome of the roll of a die on the basis of a hypothesis about the frequency of outcomes in the long run. The hypothesis conditions our expectation about how the die will come to lie. After the die has been thrown we determine the outcome by looking at it. This determination is in no way conditioned by the hypothesis concerning the frequency of outcomes in a long run of throws. Now suppose we roll the die under a shelf that prevents us from seeing how it comes to lie. We are permitted to bet after the die has been rolled but before the shelf has been removed to reveal its face. We would make the same bets, supported by the same hypothesis, that we would make in a conventional game of dice. But suppose that after a roll the player was permitted to reach under the shelf and feel the uppermost surface of the die. Those with sensitive fingers might be able to determine how the die had come to lie. Others with less sensitive fingers might be almost, but not quite, sure and be willing to express their degree of certainty with a numerical value. That numerical value would not express the frequency with which certain faces come to lie uppermost when dice are thrown. If it expressed any frequency at all, it would be the frequency with which a given person can determine by the sense of touch which face of a die is uppermost *no matter how* the die came to lie that way. At this point the frequency of outcomes in a long series of trials is irrelevant.

Thus two questions might arise in the course of a dice game. The first is, "What is the probability of rolling die so that some particular face comes to lie uppermost?" and the second is, "How is it determined which face has come to lie uppermost?" The second question seems trivial because the answer is likely to be, "By looking at it." But the determination might be based upon a retrodictive inference that did not depend upon the statistical generalization that covers the frequency of possible outcomes of dice rolling. A frequency hypothesis permits us to say of an event that has already occurred that it was one of a series in which events like it occurred with a certain frequency. But this leaves the question, "What events have actually occurred?" which is the historical question, unanswered. When someone says "He probably threw an ace," it may be supposed that he is expressing his degree of confidence, or rational credibility, in the hypothesis, "an ace was thrown." The numerical expression of the degree of confidence in the hypothesis would correspond to the numerical value for the frequency with which aces are thrown in fair games *only* if the framer of the hypothesis had no independent evidence bearing on the outcome, which is to say, only if he were unable to make a retrodiction.

But for the geologist the critical question concerns uncertainty that arises *within* the context of a retrodictive inference. Consider the case of a body in free fall. If all we knew was the instantaneous velocity with which the body struck the ground, we would have no basis for choosing any particular set of values for the initial conditions from among the infinite number of values permitted by the equation for uniformly accelerated motion. Thus no retrodiction of initial conditions is possible.

What is the difference between this case in which a retrodiction is not possible and the case of the fossil mudcracks in which a retrodiction is possible? The answer is that in the first case the initial conditions leave no *trace*, while in the second case they do. Perhaps this answer is too simple. A more general treatment which considers the further question of *why* some initial conditions leave traces and others do not may be called for. Grunbaum (1963) has treated this further question in terms of the entropy statistics of branch systems. Without going into technical details, suffice it to say that on Grunbaum's account traces, or "post-indicators," of past conditions result from the interaction of *systems.* This analysis leads him to formulate a *principle of the temporal asymmetry of the recordability of interactions.* He says (p. 283), "Hence, the exceptions apart,[1] we arrive at the fundamental asymmetry of recordability: *reliable indicators in interacting systems permit only retrodictive inferences concerning the interactions for which they vouch but no predictive inferences pertaining to the corresponding later interactions.*" This principle, or something close to it, can be formulated in everyday terms. Past events may leave traces; future events do not, at least not in the same sense or to the same degree. The reason the initial conditions cannot be retrodicted in the case of the body striking the ground lies not in the logical form of the equation that covers the event, but in the fact that the initial conditions have left no trace. Generalizations, which are intended to be invoked in retrodictions, will point to strong interactions between systems. If an interaction leaves no trace, or if the trace of an interaction is destroyed we cannot retrodict the interaction. Uncertainty about interactions comes out of uncertainty about traces.

*B*

Predictive and retrodictive uncertainty arise out of different and to some extent unrelated circumstances. Predictive uncertainty results when a prediction is based upon a general hypothesis to the effect that identical kinds of initial conditions can lead to different kinds of outcomes. Retrodictive uncertainty, on the other hand, results when a retrodiction is based upon a general hypothesis to the effect that a single kind of outcome can result from different kinds of initial conditions. The smoldering ruin of a house is a trace consistent with a great number of possible initial conditions, including faulty wiring, the leakage of natural gas, and the deliberate setting of a fire by an arsonist. But for a fire marshal, "smoldering house" is a term that includes different conditions each one of which is consistent with only one set of antecedents. By detailed specification of the outcome he may be able to adduce a hypothesis connecting this outcome with some particular antecedents.

How might uncertainty about fossil mudcracks arise? Suppose a geologist were to encounter some structures and say, "These are probably fossil mudcracks." This expression of uncertainty would not be about how mudcracks are formed or about the frequency with which, once having formed, traces of mudcracks are preserved. The geologist is uncertain about *whether* mudcracks have formed. If he knew that he was dealing with preserved mudcracks, then he could retrodict with certainty the necessary antecedents for the preservation of mudcracks. Retrodictive uncertainty arises in this case from the fact *that as far as the geologist can tell* the structure before him is consistent with more than one set of antecedents. Similarly a physician might be uncertain whether his diagnostic techniques were sensitive enough to identify paresis. *As far as he can tell*, the condition before him is consistent with a history of syphilis and consistent with a history of something else.

It may seem contrived to count these as cases of retrodictive uncertainty arising from the presumption that different antecedents are consistent with a single consequence. What are the covert hypotheses that might serve to justify such a presumption? The answer is that there are no such hypotheses, at least not in the sense that there are identifiable hypotheses that serve to justify the uncertainty of predictions about radioactive disintegration and dice rolling. Geologists are apparently more inclined to suppose that the same antecedents have different consequences than to suppose the contrary. A geologist will maintain that it is not *really* true that different initial conditions may result in indistinguishable subsequent states; it only *seems* to be true. The inability to distinguish among possible consequences is not the sort of uncertainty geologists are prepared to perpetuate in explicitly formulated hypotheses. It is not the sort of uncertainty to be enshrined metaphysically or even theoretically. It is an ephemeral uncertainty to be removed in the course of continuing investigation. The methodological consequence of this conviction among geologists is that they make no effort to formulate statistical hypotheses that might serve as grounds for assigning a numerical value to the degree of rational credibility, or probability, of a retrodictive inference, even though there is no reason in principle why they should not do so.

An example of this attitude may be found in the problem of frosted sand grains. Pettijohn (1957, p. 70) noted,

> Rarely do quartz grains show a high polish. Some sand grains, on the other hand, have a striking surface character variously described as "mat," "frosted," or "groundglass." This surface character is most commonly seen on the grains of highly quartzose and well-rounded sands of which the St. Peter (Ordovician) is the best example in the United States. Frosting has been commonly attributed to aeolian action and has been mapped in the European Pleistocene deposits by Callieux (1942), who considered the feature a criterion of periglacial wind action. The similarity of the surface to that produced on glass by sandblast gives credence to this theory, though there is little or no field evidence to support this concept. Glass, subject to the action of hydrofluoric acid, however, also acquires a frosted surface, and perhaps therefore this type of surface is a product of prolonged action by natural solvents.

Geologists did not take any of this to indicate that a single kind of surface texture could be produced under different circumstances. They proceeded to examine surface textures more carefully in the hope of distinguishing the trace produced by chemical etching from that produced by mechanical means. It is well known that with the use of the electron microscope geologists have not only been able to distinguish textures produced by etching from those produced by attrition but to distinguish among textures produced by attrition in different environments (see for example, Krinsley and Donahue, 1968).

*C*

In isolating the problem of retrodictive inference I have not done justice to the intricacy of geological practice. Historical inferences in geology are immensely complex. They do not consist of isolated retrodictive inferences, each one invoking a single generalization. Consider the relatively simple and straightforward example from Schoff (1951, p. 641), "The first orogeny affecting the Cedar Hills is inferred from the coarse conglomerates of the Indianola, which indicates large active streams with high gradients and suggests that erosion had been accelerated by the folding or uplifting of mountains." We are presented with a chain of events, including at least the uplifting of mountains, the steepening of gradients, the acceleration of erosion, the transportation of gravel, the deposition of gravel, the preservation of the sediment, and the formation of the conglomerate. Each step in this *genetic* series must, upon request for justification, be supported by appropriate generalizations.

Another complication arises from the fact that independent support for the antecedents cited in each retrodictive inference is sought. Whether or not this support is found will bear upon the credibility of the inference. A geologist is much more likely to account for some problematic scratches on a boulder as having resulted from glacial action if he has independent evidence for glaciation in the region where the boulder was found, just as a physician is more inclined to label some problematic symptoms as paresis if he knows on independent grounds that his patient has a history of syphilis. The connection between a retrodicted event and an event cited as providing independent support is often, if not predictive in the sense that it points to the future, at least directed from earlier to later time, and therefore may be supported by the same generalizations that might serve to support predictions.

Finally it should be mentioned that geologists are likely to regard geologic knowledge as incomplete unless their historical account can be understood, or explained, within the context of some comprehensive physical theory. A retrodiction does not always result in an explanation, although on occasion it may. Our inability to predict the path that a drop of water will take down a newly exposed surface is irrelevant to the practice of geology, but being able to treat the development of a drainage system as a stochastic process contributes to our understanding of the history of the earth.

## CONCLUSION

Geological knowledge presents us with a paradox. Geologists have a level of confidence in the assertions they make about the past that often approaches certainty, and yet when we examine the principles that might serve to justify these historical statements, we find that they almost always express a degree of uncertainty. The paradox is resolved by a recognition of the asymmetry of retrodictability and predictability. Geologists are not much interested in the future. They are preoccupied with what *has* happened, and they can infer, without much difficulty, some of the antecedent conditions necessary for the occurrence of events in the present. To judge geological generalizations by their ability to support predictions is absurd. Geological generalizations are commonly used in support of retrodictive inferences, and when judged by their ability to do this they measure up very well. I have expressed in rather formal terms the truism that the future is more uncertain than the past. Geologists know this very well, and so do ordinary men. It would not be necessary to belabor the point were it not for the fact that a few geologists in their discussions of uncertainty have wholly ignored retrodiction and have thereby been led to overestimate the prevalence of uncertainty in geologic knowledge. Mann's (1970) work would serve best as an analysis of a predictive historical science. But geology is not a predictive historical science. It is not even an immature predictive historical science. It is the most highly developed retrodictive historical science.

Despite the asymmetry of recordability, retrodictive uncertainty arises in geology. Its precise character has been overlooked because the general hypotheses that entail retrodictive uncertainty are hardly ever explicitly formulated. They are not so formulated, because geologists hold that the uncertainty they would express can in principle, and often in fact, be eliminated.

I have considered the problem of certainty and uncertainty in the hope of illuminating some inferential procedures in geology. I do not believe that anything I have said here has any bearing whatever on the question of determinism or the "intrinsic randomness of nature."

## References Cited


Bateman, A. M., 1950, Economic mineral deposits, 2d ed.: New York, John Wiley & Sons, Inc., 916 p.

Bagnold, R. A., 1941, The physics of blown sand and desert dunes: London, Methuen and Company, Ltd., 265 p.

Bradley, W. H., 1963, Geological laws, *in* Albritton, C. C., ed., The fabric of geology: Reading, Mass., Addison-Wesley, p. 24-48.

Chamberlin, T. C., 1904, The methods of the earth sciences: Popular Sci. Monthly, v. 66, p. 66-75.

Dunbar, C. O., and Rodgers, John, 1957, Principles of stratigraphy: New York, John Wiley & Sons, Inc., 356 p.

Flint, R. F., 1957, Glacial and Pleistocene geology: New York, John Wiley & Sons, Inc., 553 p.

Grunbaum, A., 1963, Philosophical problems of space and time: New York, Alfred A. Knopf, 448 p.

Hempel, C. G., 1965, Aspects of scientific explanation and other essays in the philosophy of science: New York, Free Press, 505 p.

Hempel, C. G., and Oppenheim, P., 1948, Studies in the logic of explanation: Philos. Sci., v. 15, p. 135-175.

Kitts, D. B., 1963, The theory of geology, *in* Albritton, C. C. ed., The fabric of geology: Reading, Mass., Addison-Wesley, p. 49-68.

Knopoff, L., 1964a, Earth tides as a triggering mechanism for earthquakes: Seismol. Soc. America Bull., v. 54, p. 1865-1870.

__________, 1964b, The statistics of earthquakes in Southern California: Seismol. Soc. America Bull., v. 54, p. 1871-1873.

Krauskopf, K. B., 1968, A tale of ten plutons: Geol. Soc. America Bull., v. 79, p. 1-18.

Krinsley, D. H., and Donahue, J., 1968, Environmental interpretation of sand grain surface textures by electron microscopy: Geol. Soc. America Bull., v. 79, p. 743-748.

LAPLACE, [P. S.], 1840, Essai philosophique sur les probabilites: Bruxelles, Soc. Belge de librairie, 265 p.
LEOPOLD, L. B., and LANGBEIN, W. B., 1963, Association and indeterminacy in geomorphology, *in* Albritton, C. C. ed., The fabric of geology: Reading, Mass., Addison-Wesley, p. 184-192.
MANN, C. J., 1970, Randomness in nature: Geol. Soc. America Bull., v. 81, p. 95-104.
NAGEL, E., 1961, The structures of science: New York, Harcourt, Brace and World, Inc., 618 p.
PEARSON, K., 1900, The grammar of science, 2d ed.: London, Adam and Charles Black, 548 p.
PETTIJOHN, F. J., 1957, Sedimentary rocks, 2d ed.: New York, Harper and Brothers, 718 p.
POPPER, K. R., 1957, The poverty of historicism: Boston, Beacon Press, 166 p.
POTTER, P. E., and BLAKELY, R. F., 1968, Random processes and lithological transitions: Jour. Geology, v. 76, p. 159-170.
ROBINSON, P. T., 1972, Petrology of the potassic Silver Peak volcanic center, Western Nevada: Geol. Soc. America Bull., v. 83, p. 1693-1708.
SCHOFF, S. L., 1951, Geology of the Cedar Hills, Utah: Geol. Soc. America Bull., v. 62, p. 619-646.
SHROCK, R. R., 1948, Sequence in layered rocks: New York, McGraw-Hill Book Company, Inc., 507 p.
SCRIVEN, M., 1959a, Truisms as grounds for historical explanations, *in* Gardener, P. ed., Theories of history: Glencoe, Ill., The Free Press, p. 443-475.
————, 1959b, Explanation and prediction in evolutionary theory: Science, v. 130, p. 477-482.
WELLER, J. M., 1960, Stratigraphic principles and practice: New York, Harper and Brothers, 725 p.

CHAPTER THREE

# The Theory of Geology

GEOLOGISTS HAVE, throughout the history of their discipline, asked questions about the nature of geology and its relationship to the other natural sciences. Self-consciousness has been increasing among geologists during the past decade as it has been increasing among all scientists. In print the most obvious sign of an inclination toward self-examination is to be found in discussions of the ideal geologic curriculum in colleges and universities. Many of the questions raised in these discussions may be interpreted as questions about the theoretical structure of the science.

## GEOLOGIC GENERALIZATIONS

In the discussion that follows, "geologic term" will be understood to mean a term which fulfills a particular function in geology and is not a term necessary for meaningful discourse outside geology. Geologic terms may have been invented for the specific role which they fill, for example, "syncline"; or they may be terms of the common language whose meaning has been expanded, or more often restricted, to fill a particular technical role, for example, "sand." A general statement which employs geologic terms in the above sense will be called a "geologic generalization."

Geologic explanations, I have suggested above (see pp. 3-27), are justified in terms of generalizations which may be compared to the laws of the other natural sciences. Most geologists have been reluctant to attach the label "law" to any of the statements which they employ. A few geologists, for example Bucher (1933), have used the term freely. It is not my intention to discuss the question of which statements should or should not be called laws. This question has been discussed at some length by most philosophers of science. For excellent discussions of the concept of a scientific law, I refer the reader to Nagel (1961), Braithwaite (1953), and Hempel and Oppenheim (1948). Although I shall have occasion to mention current opinion on the kinds of statements which are to be called laws, my main purpose here is to examine that great variety of general statements, whatever we choose to call them, which serve as a part of the logical justification for various kinds of geologic explanations.

Traditionally, scientific explanation has been regarded as a deductive operation. A generalization which functions in a deductive argument must be of strictly universal form, that is, it must admit of no exceptions whatever. Consequently, many writers have restricted the term "natural law" to statements of strictly universal form. One can cite examples of geologic generalizations which are universal. For example, the statement, "If the supply of new sand is less than that moved over the crest . . . the windward slope will be degraded as the leeward slope grows, and thus the dune will migrate with the wind by transfer of sand from one slope to the other" (Dunbar and Rodgers, 1957, p. 20) is in form, though perhaps not in principle, universal.

Generalizations employing terms denoting probability or possibility, however, far outnumber generalizations of universal form in the geologic literature. A striking feature of geologic discourse is the frequency with which such words and phrases as "probably," "frequently," and "tends to," occur in generalizations.

Explanations which contain these nonuniversal general statements are not deductive, but rather inductive in form. As Hempel (1958, p. 40) has put it, "This kind of argument, therefore, is inductive rather than strictly deductive in character: it calls for the acceptance of *E* [a sentence stating whatever is being explained] on the basis of other statements which constitute only partial, if strongly supporting, grounds for it." Few geologists would disagree that the generalizations which they employ seldom provide more than "partial, if strongly supporting, grounds" for their conclusions.

It is customary to call general statements which admit of some exceptions "statistical" or "probabilistic." According to Bunge (1961, p. 267) statistical lawlike statements are ". . . propositions denoting regularities in the large or in the long run. They contain logical constructs belonging to mathematical statistics and characterizing central or overall trends ('average,' 'dispersion,' 'correlation,' and the like)." The generalization which states the rate of decay of carbon fourteen is statistical in this sense.

General statements which contain terms denoting probability or possibility but which are characterized by no precise formulation of statistical constructs in numerical terms are better labeled "probabilistic" than "statistical." The geologic literature abounds in statements of this sort. For example: "Wind-driven sand, like snow, tends to settle in drifts in the wind-shadow of topographic obstructions" (Dunbar and Rodgers, 1957, p. 19).

Statistical inference is playing an increasingly important role in geologic investigation and one cannot help being impressed by the frequency with which statistical constructs appear in geologic literature. In much of this literature, however, the precise statistical terminology employed may obscure the loosely probabilistic character of the generalizations containing it. In the following quotation the antecedent portions of the generalization are framed with statistical precision but the operators, for example "most" and "many," lack such precision.

> Dune sands *tend to be* better sorted than river sands and a plot of standard (sorting) against mean grain-size indicates three fields, one for river sands, one for dune sands, and a third field of overlap. This figure points out that *many* river sands can be distinguished from dune sands and vice versa on the basis of their textural parameters but that a wide field of overlap exists. In practice this field of overlap is not necessarily a serious matter, since *most* coastal, barrier bar, and lake dune sands have a standard deviation of less than 0.40, and *many* desert and inland dune sands do not exceed 0.50, whereas *most* river sands have a standard deviation in excess of 0.50. (Friedman, 1961, p. 524, italics mine)

Scriven (1959) has recently suggested that the generalizations in terms of which explanations of individual events are justified are usually neither universal nor statistical, nor even probabilistic, but rather belong to another category of statements which he calls "normic statements." Scriven's concept of the normic statement throws particular light on the problem of geologic generalizations and explanations. He states (p. 464), "I suggest there is a category of general statements, a hybrid with some universal features and some statistical features, from which alone can be selected the role-justifying grounds for good explanations of individual events," and (p. 466),

> The statistical statement is less informative than the normic statement in a very important way, although an exact statistical statement may be informative in a way a normic statement is not. The statistical statement does not say anything about the things to which it refers except that some do and some do not fall into a certain category. The normic statement says that *everything* falls into a certain category *except* those to which *certain special conditions* apply. And although the normic statement itself does not explicitly list what counts as exceptional conditions, it employs a vocabulary which reminds us of our knowledge of this, our trained judgment of exceptions.

And (p. 466),

> Now if the exceptions were few in number and readily described, one

could convert a normic statement into an exact generalization by listing them. Normic statements are useful where the system of exceptions, although perfectly comprehensible in the sense that one can learn how to judge their relevance, is exceedingly complex. We see immediately the analogy with—in fact, the normic character of—physical laws. The physicist's *training* makes him aware of the system of exceptions and inaccuracies, which, if simpler, could be explicitly in the statement of scope.

And finally (p. 467),

. . . statistical statements are too weak—they abandon the hold on the individual case. The normic statement tells one what had to happen in *this* case, unless certain exceptional circumstances obtained; and the historical judgment is made (and open to verification) that these circumstances did not obtain.

Let us consider again the generalization from Dunbar and Rodgers (1957, p. 19) concerning the accumulation of wind-driven sand. "Wind-driven sand, like snow, tends to settle in drifts in the wind-shadow of topographic obstructions." Offhand one would be inclined to regard this statement as probabilistic because of the distinct probabilistic, or even statistical, connotation of the phrase "tends to." But was it the intention of the authors to assert that if certain specified conditions are realized then *in a majority of cases* certain other specified conditions will follow, in which case the generalization must indeed be regarded as probabilistic; or was it their intention to assert that if certain specified conditions are realized then certain other specified conditions will *always* follow *except* where certain special conditions apply, in which case the generalization may be regarded as normic? I do not know what the intentions of the authors were, but in the absence of this knowledge it seems to me fully as plausible to regard the statement as normic as to regard it as probabilistic. The question of meaning raised in connection with this generalization can be raised in connection with a great number of seemingly probabilistic geologic generalizations, and, for that matter, about a great number of seemingly universal statements.

Most geologic generalizations, whatever their explicit form, could be regarded as normic statements, and the sense in which we actually understand them is better conveyed by the term "normic" than by the terms "universal," "statistical," or "probabilistic." One thing is certain and that is the geologist will not "abandon the hold on the individual case" by allowing statistical statements to assume too important a role in his procedures, for, as I have suggested earlier, it is the individual case which is his primary concern.

The normic character of some generalizations is made explicit.

Whenever the terms "normally" or "ordinarily" are encountered in a generalization it can usually be assumed that the statement is normic. The following statement is explicitly normic. "Carbonaceous material is a residue that remains after various more or less mobile compounds, produced during the decomposition of organic matter in an oxygen-deficient environment, have moved away. Methane gas, for example, evolves and escapes under ordinary conditions" (Weller, 1960, p. 152).

Scriven holds out the hope at least that normic statements can be converted into universal statements. Von Engeln's (1942, p. 457) statement of the "law of adjusted cross sections" represents an attempt to list "exceptional circumstances" which, if they should apply, invalidate the generalization.

> Given that the surfaces of the joining glaciers are at the same level, that the width of the main valley is not abruptly increased below the junction point, and that the rate of motion of the main glacier is not greater below than above the junction—and all these conditions are met in numerous instances—it follows that there must be an abrupt increase in depth of the main valley to accommodate the volume of the combined ice streams at the place where they join.

It is clear that geologists tolerate a good deal more imprecision in their generalizations than is technically necessary. Let us turn once again to the generalization from Dunbar and Rodgers (1957, p. 19) on the accumulation of wind-driven sand. It might be possible to specify the circumstances under which wind-driven sand would *always* accumulate in the wind-shadow of topographic obstructions. To accomplish this it would be necessary to specify restricted ranges for the pertinent initial and boundary conditions such as the velocity of the wind, the character and quantity of the sand, the form of the obstruction, etc. The framing of universal statements is not, however, the primary goal of the geologist. The primary goal is to frame *general* statements, universal or not, on the basis of which explanations can be justified. The introduction of specific initial and boundary conditions may permit the formulation of strictly universal statements, but these statements will be of no use whatsoever unless these specific conditions can be independently determined. Very often it is not possible to determine these conditions, and generalizations are intentionally left in a loosely formulated state. This does not preclude the possibility of taking certain specific conditions into account when these conditions are capable of being independently inferred.

The view, not uncommon among geologists, that geology is as much an "art" as a science may stem from the fact that the system of exceptions associated with geologic generalizations is usually so complex that "judgment" of their relevance may play an important role in investigation—more important a role than in the natural sciences in which the system of exceptions seems simpler.

The employment of many seemingly statistical and probabilistic generalizations imparts to geology an aspect of "indeterminacy." It would be very misleading to equate, as some geologists have done, this geologic indeterminacy with the indeterminacy principle of modern physics. In modern quantum-statistical mechanics indeterminacy is an integral part of theory. It is *not* assumed that uncertainty can be removed by a more complete and detailed specification of a particular set of variables. In geology, it seems to me, the opposite assumption is usually made. Uncertainty is regarded as a feature to be tolerated until more complete knowledge of variables allows its replacement with certainty or, to put it another way, until probabilistic and normic statements can be replaced by universal statements. Geologists are fundamentally deterministic in their approach to scientific investigation. Indeterminacy, in the sense that it is understood in physics, may enter into geology in those cases where the principles of statistical mechanics enter directly into a geologic inference.

One final point should be made in connection with probability. Logicians have called attention to the fact that there are two senses in which the term "probability" may be used in connection with scientific hypotheses. In the first sense it may be asserted that a particular hypothesis is probable or more probable than some other hypothesis; for example, "It is probable that glauconite marl represents the normal primary occurrence and that greensand is a concentrate brought together like any other kind of sand during transportation on the sea floor" (Dunbar and Rodgers, 1957, p. 184). In the second sense a probability is assigned to an event within a hypothesis, for example, "Almost all moving masses that begin as landslides in subaqueous situations become mud flows" (Weller, 1960, p. 158). Probabilistic terms are so frequently and loosely employed by geologists that it is often difficult to determine in which of the two senses a particular probabilistic term is to be understood.[1]

Whether or not a geologic generalization is of universal form is certainly not the only question which need concern us. A statement may be of strictly universal form but be restricted in scope. The scope of a statement corresponds to the class of objects or events to which the statement applies. Every statement is to some degree restricted in scope. Certain kinds of restrictions of scope can seriously detract from the usefulness of a general statement which is intended to have explanatory power. Among the most serious of these restrictions is that imposed by reference to particular objects, or to finite classes of objects, in particular spatio-temporal regions.

Generalizations which contain references to particular objects, or which contain terms whose definition requires reference to particular objects, are rare in geology, as they are in other sciences. Generalizations in which reference is made to a particular finite class of objects falling into a definite spatio-temporal region are quite common and could conceivably cause some difficulty. In the statement, "Major streams of the northern Appalachian region rise in the Allegheny Plateau and flow southeastward to the Atlantic directly across the northeast-southwest grain of rocks and structures ranging in age from Precambrian to Tertiary" (Mackin, 1938, p. 27), for example, it is clear that the scope is restricted to a finite class of objects. Obviously the statement has no explanatory power. We should not be justified in explaining the fact that a particular major stream of the northern Appalachian region flowed southeastward by reference to this statement, because, as Nagel (1961, p. 63) puts it, "if a statement asserts in effect no more than what is asserted by the evidence for it, we are being slightly absurd when we employ the statement for explaining or predicting anything included in this evidence." No geologist would attempt to explain anything at all by reference to this statement, nor did Mackin intend that they should. The statement was obviously framed not to provide a means of explanation and prediction, but rather to convey, as economically as possible, information about some particular objects in a particular region. The statement could have been formulated as a conjunction of singular statements without change of meaning. Many general statements in geology whose scope is explicitly restricted are of this type.

As Nagel (1961, p. 63) has pointed out, a general statement which refers to a group of objects or events which is presumably finite may be assigned an explanatory role if the evidence for the statement is not assumed to exhaust the scope of predication of the statement. The evidence cited to support the statement, "As the velocity of a loaded stream decreases, both its competence and its capacity are reduced and it becomes *overloaded*" (Dunbar and Rodgers, 1957, p. 9), for example, would consist of a finite number of observations, and yet it is clearly not the intention of the authors to assert that this set of observations exhausts the scope of the statement.

A problem about the intention of the framer of a general statement might arise. This problem is particularly likely to come up in connection with general statements concerning the association of past events where the number of instances cited in support of the statement is small. Do the instances cited to support the following generalization exhaust its scope of predication, or not? "Fold mountains have their origin in the filling of a geosyncline chiefly with shallow water sediments, conglomerates, sandstone, shales, and occasional limestones" (von Engeln and Caster, 1952, p. 234). The class of fold mountains contains a finite number of individuals. The question of scope cannot be answered by an analysis of the statement, nor can it be answered by examining the evidence for the statement. The answer lies in a determination of the intention of the authors. Is it their intention merely to convey information about a finite number of instances, or do they intend to assert that fold mountains, past, present and future, have their origin in the filling of geosynclines? If the latter is the case, then the intention is to assign an explanatory, and possibly even a predictive, role to the statement. I cite this example to illustrate the importance of intention. I have no doubt that the authors meant to go beyond a mere historical report.

It might be argued that all general geologic statements are restricted in scope, because they contain—implicitly or explicitly—the individual name, "the planet earth." Nagel (1961) regards Kepler's laws of planetary motion as lawlike even though they contain individual names because, as he puts it (p. 59), "The planets and their orbits are not required to be located in a fixed volume of space or a given interval of time." I think it is true of geologic generalizations that the objects covered are not required to be associated with "the planet earth." The generalization will hold wherever certain, to be sure very specific, conditions obtain. In other words, it would be possible, in principle at least, to express every term, including "the planet earth," in universal terms and so eliminate essential reference to any particular object. The fact is, of course, that no one worries about essential reference to the earth, and no one need worry so long as we practice our profession on earth.

A serious problem concerns suspected but unstated restriction of the temporal scope of geologic statements. Because this problem stands at the very core of any consideration of the uniformitarian principle, I shall discuss it in a separate section.

## THE THEORY OF GEOLOGY

Is there any justification whatever for speaking about a "theory" of geology? The answer to this question obviously hinges on another, namely, "What do we mean by theory?" In an attempt to answer the latter question I shall quote three philosophers of science on the subject. Hempel (1958, p. 41) has this to say:

For a fuller discussion of this point, it will be helpful to refer to the familiar distinction between two levels of scientific systematization: the level of *empirical generalization*, and the level of *theory formation*. The early stages in the development of a scientific discipline usually belong to the former level, which is characterized by the search for laws (of universal or statistical form) which establish connections among the directly observable aspects of the subject matter under study. The more advanced stages belong to the second level, where research is aimed at comprehensive laws, in terms of hypothetical entities, which will account for the uniformities established on the first level.

Nagel (1961, p. 83) recognizes the usefulness of a distinction between "theoretical laws" and what he calls "experimental laws," but calls attention to the difficulties which may arise in the attempt to label a particular law as one or the other. As he points out, however:

> Perhaps the most striking single feature setting off experimental laws from theories is that each "descriptive" (i.e. nonlogical) constant term in the former, but in general not each such term in the latter, is associated with at least one overt procedure for predicting the term of some observationally identifiable trait when certain specified circumstances are realized. The procedure associated with a term in an experimental law thus fixes a definite, even if only partial, meaning for the term. In consequence, an experimental law, unlike a theoretical statement, invariably possesses a determinate empirical content which in principle can always be controlled by observational evidence obtained by those procedures.

And finally Carnap (1956, p. 38), emphasizing the linguistic consequences of the distinction made by Nagel and Hempel, states:

> In discussions on the methodology of science, it is customary and useful to divide the language of science into two parts, the observation language and the theoretical language. The observation language uses terms designating observable properties and relations for the description of observable things or events. The theoretical language, on the other hand, contains terms which may refer to unobservable events, unobservable aspects or features of events, e.g., to microparticles like electrons or atoms, to the electromagnetic field or the gravitational field in physics, to drives and potentials of various kinds in psychology, etc.

We may now proceed to examine geology with these "theoretical" characteristics in mind. The claim has been repeatedly made that geology was in the past, and to some extent remains today, "descriptive." For example, Leet and Judson (1954, p. ii) state, "Originally geology was essentially descriptive, a branch of natural history. But by the middle of the 20th century, it had developed into a full-fledged physical science making liberal use of chemistry, physics and mathematics and in turn contributing to their growth." Just what is meant by "descriptive" in this sort of statement? It certainly cannot be taken to mean that geology is today, or has been at any time during the last two hundred years, wholly, or even largely, concerned with the mere reporting of observations. The very fact that during this period geologists have remained historical in their point of view belies any such contention. The formulation of historical statements requires inferential procedures that clearly go beyond a mere "description."

The feature that has apparently been recognized by those who characterize geology as descriptive, and indeed the feature which we can all recognize, is that most geologic terms are either framed in the observation language or can be completely eliminated from any geologically significant statement in favor of terms that are so framed. This is simply to say that most geologic terms are not theoretical. Even Steno's four great "axioms" may be regarded as "observational" rather than "theoretical."

The abundance of historical terms in geology may give rise to some confusion when an attempt is made to distinguish between observation terms and theoretical terms. Historical terms involve some kind of historical inference to a past event or condition and consequently require for their definition reference to things which have not been observed by us. Consider, for example, the term "normal fault" which has been defined as a fault "in which the hanging wall has apparently gone down relative to the footwall" (Billings, 1954, p. 143). The label "theoretical" would probably be denied this term by most philosophers of science because the past event can be described in the observation language and could consequently, in *principle* at least, be observed.

The question then arises, are there *any* geologic terms which clearly qualify as theoretical? This is a difficult question because, as many authors have pointed out, the distinction between theoretical terms and observation terms is far from clear. There are certainly no geologic terms so abstract as the higher-level theoretical constructs of quantum mechanics, for example. On the other hand, there are a number of geologic terms which almost certainly qualify as what might be called "lower-level theoretical terms," that is, terms of relatively limited extension. "Geosyncline," it seems to me, is such a term. According to Kay (1951, p. 4) a geosyncline may be defined as "a surface of regional extent subsiding through a long time while contained sedimentary and volcanic rocks are accumulating; great thickness of these rocks is almost invariably the evidence of the subsidence, but not a necessary requisite. Geosynclines are prevalently linear, but nonlinear depressions can have properties that are essentially geosynclinal."

Another candidate for the title "theoretical term" is "graded stream." According to Mackin (1948, p. 471),

> A graded stream is one in which, over a period of years, slope is delicately adjusted to provide, with available discharge and with prevailing channel characteristics, just the velocity required for the transportation of the load supplied from the drainage basin. The graded stream is a system in equilibrium; its diagnostic characteristic is that any change in any of the controlling factors will cause a displacement of the equilibrium in a direction that will absorb the effect of the change.

If I am correct in regarding these terms as theoretical, we might expect that some difficulties would arise if an attempt were made to define these expressions in terms of observables. I do not, however, feel inclined or qualified to go into the problem of "rules of correspondence" at this point.

What is the function of theoretical terms in a scientific system? The answer to this question was given, I think, in the passage from Hempel (1958, p. 41) quoted earlier, and in the words of Nagel (1961, pp. 88-89) when he says,

> An experimental law is, without exception, formulated in a single statement; a theory is, almost without exception, a system of several related statements. But this obvious difference is only an indication of something more impressive and significant: the greater generality of theories and their relatively more inclusive explanatory power.

The general statements which contain the terms "geosyncline" and "graded stream" have, relative to other geologic statements, great generality and quite clearly fulfill the function which theoretical terms are designed to fill; that is, they establish some considerable "explanatory and predictive order among the bewildering complex 'data' of our experience, the phenomena that can be directly observed by us" (Hempel, 1958, p. 41).

One of the most striking features of scientific systems with a highly developed theoretical structure is the degree to which it is possible to make logical connections between the various general statements contained in the system. A given general statement may, in combination with other general statements or even with singular statements, serve as the basis for the deductive derivation of other general statements, or itself be derivable from other generalizations in the same way. It is this feature which provides empirical systems with their systemicity. The importance of system is indicated by Braithwaite (1953, pp. 301-2), who holds that a hypothesis is to be considered "lawlike" only on the condition that it "either occurs in an established scientific deductive system as a higher-level hypothesis containing theoretical concepts or that it occurs in an established scientific deductive system as a deduction from higher-level hypotheses which are supported by empirical evidence which is not direct evidence for itself."

As Braithwaite suggests, in the case of generalization (*G*) which is deductively related to other generalizations, it is possible to distinguish between indirect and direct supporting evidence for it. Any empirical evidence which supports a generalization which is deductively related to *G* will count as indirect evidence in support of *G*. Nagel (1961, p. 66) emphasizes the importance of indirect evidence in the statement, "Indeed, there is often a strong disinclination to call a universal conditional *L* a 'law of nature,' despite the fact that it satisfies the various conditions already discussed, if the only available evidence for *L* is direct evidence."

How much systematization is there in the body of geologic knowledge? If we consider geologic generalizations, that is, general statements containing only special geologic terms, we are immediately struck with how little systematization obtains. There is some, of course, and it is usually provided in terms of such unifying concepts as "graded steam" and "geosyncline." This low-degree systematization is not surprising, for, as every elementary geology text insists, geology is not a discipline unto itself. We have only to introduce into our system the generalizations and laws of physical-chemical theory and the logical connections within the branches and between the branches of geology become more impressive. It may be a matter of contention among geologists as to whether every geologic generalization has been, or in principle could be, incorporated into a theoretical system in terms of physical-chemical laws. If a geologic generalization is truly independent, however, no matter how useful it may be, its status will suffer because no indirect evidence can be presented to support it, and we are likely to speak of it as "merely" an empirical generalization.

The question is not whether geology is chemistry and physics, but whether or not geologists will utilize physical-chemistry theory with all its admitted imperfections and all its immense power. The question has been answered. Almost all geologists proceed as if every geologic statement either has now, or eventually will have, its roots in physical-chemical theory. Geologic theory is modified to accommodate physical-chemical theory and never conversely. Furthermore, this confidence in physics and chemistry is not a unique feature of twentieth-century geology, for Hutton (1795, pp. 31-32) said:

> It must be evident, that nothing but the most general acquaintance with the laws of acting substances, and with those of bodies changing by the powers of nature, can enable us to set about this undertaking with any

reasonable prospect of success; and here the science of Chemistry must be brought particularly to our aid; for this science, having for its object the changes produced upon the sensible qualities, as they are called, of bodies, by its means we may be enabled to judge of that which is possible according to the laws of nature, and of that which, in like manner, we must consider as impossible.

Certainly no consistent, economical, complete deductive system of geology exists, but I think that we can detect a suggestion of such a system. In this system the higher-level universals, or postulates, which serve as the basis for the deductive derivation of other generalizations and of singular statements of the system are exactly those universals which serve this purpose in physics and chemistry. The theory of geology is, according to this view, the theory of physics and chemistry. The geologist, however, unlike the chemist and the physicist, regards this theory as an instrument of historical inference.

The generalizations containing the theoretical terms "geosyncline" and "graded stream" are not regarded as "fundamental laws of nature" or as postulates of a theory of geology. Geologists are committed to the view that generalizations containing these terms can be derived from higher-level generalizations which belong to physics. Thus Leopold and Langbein (1962, p. A 11) state with reference to the graded-stream state, "A contribution made by the entropy concept is that the 'equilibrium profile' of the graded river is the profile of maximum probability and the one in which entropy is equally distributed, constituting a kind of isentropic curve."

The activity of attempting to relate geologic generalizations to physical-chemical theory is not a game played by geologists for its own sake, nor is it played solely for the sake of increasing our confidence in these generalizations by providing indirect supporting evidence for them. It is the hope of those engaged in this activity, which I am tempted to call "theoretical geology," that putting a geologic generalization into a theoretical setting will allow a more precise and rigorous formulation of it, and thus may increase its utility. This consideration is undoubtedly behind Leopold and Langbein's statement (1962, p. A 1), "The present paper is an attempt to apply another law of physics to the subject for the purpose of obtaining some additional insight into energy distributions and their relation to changes of land forms in space and time."

A major objection to the view that geology is "descriptive" and thus in some sense not "mature" is that the scope of geologic knowledge and theory is in no sense now, nor has it ever been, exhausted by what can be framed in special geologic terms and in terms of the common language alone. The laws which provide for much of the higher-level systematization in geology are perhaps not immediately associated with a theory of geology because they are laws without particular geologic reference which are ordinarily recognized as belonging to the theory of chemistry and physics. To consider the science of geology as composed of just those statements containing particular geologic references amounts to a wholly artificial decapitation of a rather highly organized theoretical structure. An arbitrary distinction among physical, chemical, and geologic terms is not significant here. What is significant is the role played by laws and terms, no matter what their origin or what we may choose to call them, in the theory of geology. Modern geology assumes *all* of contemporary physical-chemical theory and presents on the basis of this assumption a high degree of logical integration.

It has been suggested, for example by Leet and Judson (1954, p. ii) in the statement quoted above, that geology not only draws upon physical-chemical theory but may contribute to this theory. In some sense, at least, this is true since geologic problems may serve to stimulate investigations which are primarily physical or chemical in nature, particularly if the problems arise in such fields as geochemistry, geophysics, and crystallography in which the borderline between geology and the rest of physical science is difficult to draw. Traditionally, however, geologists have not tampered with nongeologic theory in any very direct way. The physical-chemical theory of their time is a standard which tends to be accepted by each generation of geologists.

## THE ORIGIN OF GEOLOGIC GENERALIZATIONS

It has often been said that geology is "inductive." It may be that, in some cases at least, the application of this label is meant to convey the fact that the explanations proposed by geologists are probabilistic rather than deductive. Usually, however, I think that the term "inductive" when applied to geology is meant to convey that the generalizations of geology are to be regarded as inductive generalizations. If we agree that most geologic generalizations are not theoretical, then this latter view of the inductive nature of geology is entirely plausible since, as Nagel (1961, p. 85) has pointed out, "An immediate corollary to the difference between experimental laws and theories just discussed is that while the former could, in principle, be proposed and asserted as inductive generalizations based on relations found to hold in observed data, this can never be the case of the latter." The problem of the process by which theoretical laws are formulated is far beyond the scope of this chapter.

Although the usual procedure in geology is to explain a generalization which has been formulated inductively in terms of higher-level generalizations, using the laws of chemistry and physics, it is possible to start with these higher-level generalizations and deduce from them generalizations of geologic significance and utility. We thus speak of the deduction of the mineralogical phase rule from the phase rule of Gibbs (see, for example, Turner, 1948, p. 45). It is my belief that the most dramatic advances in geologic theory are to be expected from this deductive method.[2]

## THE UNIFORMITARIAN PRINCIPLE

There is widespread agreement among geologists that some special principle of uniformity is a fundamental ingredient of all geologic inference. Longwell and Flint (1955, p. 385) go so far as to say, "The whole mental process involved in this reconstruction of an ancient history is based on that cornerstone of geologic philosophy, the principle of uniformitarianism, probably the greatest single contribution geologists have made to scientfic thought." Despite this general agreement about the importance of the principle, geologists hold widely divergent views as to its meaning. So divergent are these views, in fact, that one is led to conclude that there has been little or no resolution of the problems which gave rise to the famous controversies between the "uniformitarians" and the "catastrophists" in the nineteenth century. Though the problems have not been solved, the controversy has subsided. A number of accounts of the history of the uniformitarian principle are available, including an excellent one by Hooykaas (1959).

There are two principal problems concerning the concept of uniformity. The first problem concerns the grounds upon which the assertion of uniformity rests. The second problem concerns the precise nature of the restriction which it imposes.

As to the grounds upon which the assertion of uniformity rests, it would be difficult, if not impossible, to contend that any meaningful empirical support can be presented for the view that "geologic process" in the past was in any way like this process in the present, simply because statements about process must be tested in terms of statements about particulars and no observation of past particulars can be made. We do, of course, formulate statements about particular past conditions. These statements are not supported by direct observations, however, but rather are supported by inferences in which some assumption of uniformity is implicit. *In terms of the way a geologist operates, there is no past until after the assumption of uniformity has been made.* To make statements about the past without some initial assumption of uniformity amounts, in effect, to allowing any statement at all to be made about the past. Geologists, as I have suggested earlier, are primarily concerned with deriving and testing singular statements about the past. The uniformitarian principle represents, among other things, a restriction placed upon the statements that may be admitted to a geologic argument, initially at the level of primary historical inference. Once the assumption of uniformity has been made for the purposes of primary historical inference, then it may be possible to "demonstrate" some uniformity or, for that matter, depending upon how strictly the principle is applied in the first place, some lack of uniformity. The essential point is that the assumption of uniformity must precede the demonstration of uniformity and not vice versa. In principle, at least, it would be possible to make at the outset some assumptions about particular conditions at some time in the past and on the basis of these assumptions "test" whether some law held at that time. Geologists have chosen to do the opposite, however, by making some assumptions about the uniformity of the relationships expressed in a law and on this basis deriving and testing some singular statements about the past.

The view that the uniformity of "nature" or of "geologic process" represents an *assumption* that is made in order to allow geologists to proceed with the business of historical inference is a very old one. In Hutton (1788, p. 301-302), for example, we find this view expressed in the statement, "We have been representing the system of this earth as proceeding with a certain regularity, which is not perhaps in nature, but which is necessary for our clear conception of the system of nature." And Lyell (in a letter to Whewell, 1837, from "Life, Letters and Journals," 1881, Vol. II, p. 3) expresses a similar view when he states,

> The former intensity of the same or other terrestrial forces may be true; I never denied its possibility; but it is conjectural. I complained that in attempting to explain geological phenomena, the bias has always been on the wrong side; there has always been a disposition to reason *a priori* on the extraordinary violence and suddenness of changes, both in the inorganic crust of the earth, and in organic types, instead of attempting strenuously to frame theories in accordance with the ordinary operations of nature.

The uniformitarian principle is not usually formulated as an *assumption* by contemporary geologists.

If the uniformitarian principle is to be regarded as a methodological device rather than an empirical generalization, a reasonable step toward a solution of the problems surrounding the principle would be an attempt to explicate it. "What statements in the theory of geology shall be regarded as untimebound?" is a question which might be asked in connection with such an attempt.

It is plausible to regard every singular descriptive statement in geology as timebound, to some degree at least, simply because a principal aim of geologic endeavor is to "bind" singular descriptive statements temporarily, which is another way of saying that a principal aim of geologic endeavor is to produce a historical chronicle. Singular descriptive statements may apply to different times and different places, but it cannot be assumed that they apply to all times and to all places, and consequently their uniformity cannot be assumed for the purposes of inference.

There are few geologists who would disagree with this view regarding singular descriptive statements, and for this reason I may be accused of propounding the obvious. Before the accusation is made, however, let us consider a passage from Read (1957, p. 26):

> This difficulty confronting static metamorphism has been tackled by Daly, who meets it by relaxing the rigidity of the doctrine of uniformitarianism. He admits that, compared with its proposed potency in the Pre-Cambrian times, load metamorphism must have been of relatively little importance in later geological eras. To account for this, he assumes that the earth's thermal gradient was steeper during the formation of the Pre-Cambrian so that regional metamorphism under a moderate cover was possible. He considers that this speculation concerning a hotter surface to the earth is "no more dangerous than the fashionable explanation of all, or nearly all, regional metamorphism by orogenic movements." I agree that. . . [although] uniformitarianism suits the events of the 500 million years of geological history as recorded in the Cambrian and later fossiliferous rocks, it may quite likely not be so valid for the 2,000 million years of Pre-Cambrian time.

To suggest that the thermal gradient of the earth was steeper during the Pre-cambrian than it is today is to make an assertion about particular conditions during the past. This assertion is perhaps not so soundly based as many others, but it has resulted from the kind of retrodictive inference that geologists frequently perform. To claim that this statement requires a relaxation of the doctrine of uniformitarianism is, in principle, equivalent to claiming that an assertion that the topography of Oklahoma during Permian time was different from what it is today requires a similar relaxation.

The statements that the geologist wishes to regard as untimebound are of general form. This is revealed by the fact that he speaks of uniformity of "cause" or "process" or "principle" or "law," concepts which are expressed in general statements. The whole problem of the strictness with which the uniformitarian assumption is to be applied revolves ultimately around the question of *which* of the many statements of general form in the theory of geology are to be regarded as being without temporal restriction.

Geologists usually speak of the problem of uniformitarianism as though it were a problem of whether or not the relationships expressed in scientific laws could be regarded as constant throughout geologic time. Is this really the problem? Are geologists bothered, for example, by the question of whether or not the relationship expressed in the equation $F = G \cdot m_1 m_2 / d^2$ can be regarded as independent of time? They may occasionally speculate about the "uniformity of nature," but such speculations do not enter into the conduct of geologic investigation in any significant way. Here again it is to the authority of physical-chemical theory that appeal is made. If an expression is untimebound in this theory, it is untimebound in the theory of geology, particularly if the expression is regarded as a "fundamental law." Furthermore, if a law should be formulated in which the expressed relationship was a function of the passage of historic time, and if this law should come to be regarded as a valid part of some physical or chemical theory, there would probably be no reluctance on the part of geologists to accept the law, nor would the acceptance signal an abandonment of the uniformitarian principle. The view that it is "law" which is to be regarded as uniform has been suggested many times before, for example in the following statement from Moore (1958, p. 2): "As foundation, we accept the conclusion that nature's laws are unchanging."

Not only is the most general form of a fundamental law regarded as untimebound, but so are substitution instances and other logical consequences of such a law. We should agree, for example, that at any time and place where two spherical masses of one pound each and of uniform density are held with their centers one foot apart, the force of attraction between them will be $3.18 \times 10^{-11}$ pounds weight. We do not conclude from this, however, that the attraction between *any* two objects at *any time* is $3.18 \times 10^{-11}$ pounds weight, for to do so would necessitate the assumption of the temporal uniformity of certain particular conditions. This particular attractive force is to be inferred only where the specified antecedent conditions actually obtain. That these particular conditions, or for that matter any particular conditions, did in fact obtain at a particular time and place is a matter for independent verification. The law does not "change," but different substitution instances of the law are applicable to different times and places. The immediate problem which the geologist faces is not one of uniformity, but one of determining which, if any, of the infinite number of substitution instances of a general law is applicable in a particular case.

Let us consider a geologic statement in which doubt about temporal extension is expressed. "Probably only time and the progress of future studies can tell whether we cling too tenaciously to the uniformitarian principle in our unwillingness to accept fully the rapid glacial fluctuations as evidenced by radiocarbon dating" (Horberg, 1955, p. 285). The author has suggested, has he not, that there was available a generalization concerning the rate of glacial fluctuation, perhaps supported by observations of extant glaciers, whose validity had become doubtful because of our greater confidence in another generalization. Isn't there something suspicious about a statement which is supposed to have some degree of temporal extension and which contains a reference to a particular rate or even to a limited range of rates? There are many physical laws in the form of equations which allow the calculation of a particular rate when particular values are substituted for variables. The unsubstituted form of a physical equation, however, contains no reference to a particular rate. Rate of glacial fluctuation does not remain constant throughout time any more than the force of attraction between objects remains constant throughout time. Rate of glacial advance and retreat is dependent upon a number of variables, among which are the topography, the thickness of the ice, and the temperature of the atmosphere. Unless specific values can be substituted for each of these variables, a law of glacial movement cannot be meaningfully applied at all, and could certainly not be called upon to serve as the basis for the assertion that the rate of glacial movement had some uniformity in geologic time. Horberg's problem did not involve a question of the "uniformity of nature," but rather it involved a question of whether or not he was in a position to determine particular values for each of the variables upon which rate of glacial movement might be presumed to be dependent. No explication of the concept of "the uniformity of nature" nor "universal causation" could have served as a basis for the solution of this problem.

Another problem presents itself at this point. Is there a law of glacial movement which has been formulated with sufficient precision to allow us to say what the pertinent variables are, let alone determine specific values for them? The answer would have to be that there is not. What is available is an imprecisely formulated generalization, probably normic in form, which may serve as the basis for a variety of inferences but cannot serve as a basis for a calculation of the rate of glacial advance and retreat during a particular interval of time. The temporal universality of imprecisely formulated probabilistic and normic generalizations will always be suspect simply because it is characteristic of such statements that they contain what must be regarded as hidden references to particulars. We are able to increase our confidence in the temporal universality of a generalization by formulating it with greater precision. It must be borne in mind, however, that progress in geology has depended upon the willingness of geologists to assume the temporal uniformity of a great variety of imprecisely formulated generalizations.

There is a long-standing philosophical problem concerning the uniformity of nature. It has been discussed at great length in the past and it will be discussed in the future. This problem is everybody's problem, not a special problem for geologists. The special problem for geologists concerns first the availability of appropriate laws, and second the applicability of laws to particular situations. In an attempt to solve the problem, many geologists are engaged in work that is directed toward the goal of increasing the precision with which geologic generalizations are formulated. They proceed by observing in the field, experimenting in the laboratory and, increasingly, by paper and pencil operations within a theoretical framework. The latter procedure may turn out to be the most satisfying of all because never are we so confident in the precision of a law as when it has been made a part of some theory.[3]

The precise formulation of geologic generalizations cannot be expected to lead to an immediate solution of all the problems facing the contemporary geologist. The principal use to which geologic generalizations are put is a basis for explanatory inferences which allow the derivation and testing of singular statements about the past. A generalization cannot be used in this way unless specific values can be substituted for most of the variables contained in it. A specific value which is substituted for a variable in a generalization employed in a retrodictive inference must have been determined on the basis of another retrodictive inference which is itself dependent upon some generalization. The necessary interdependence of the bewildering variety of geologic retrodictive inferences upon one another seems, at times, to result in a piling of uncertainty upon uncertainty. Remarkably, however, confidence in geologic statements about the past as reliable descriptions of particular conditions, in particular places, at particular times, is high. The basis of this confidence is that not only are retrodictive inferences to a large extent dependent upon one another, but that they serve as the only means of verification of one another. Each statement about the past is tested against other statements about the past. Laboriously we build the chronicle, selecting, eliminating, and modifying as we proceed, bringing to bear, in addition to an immensely complicated inferential apparatus, our trained judgment.

## References Cited

Billings, M. P., 1954, Structural geology: New York, Prentice-Hall, 514 p.

Braithwaite, R. B., 1953, Scientific explanation: New York, Cambridge University Press, 375 p.

Bucher, W. H., 1933, The deformation of the earth's crust; an inductive approach

to the problems of diastrophism: Princeton, Princeton University Press, 518 p.
BUNGE, M., 1961, Kinds and criteria of scientific laws: Phil. Sci., vol. 28, pp. 260-281.
CARNAP, R., 1956, The methodological character of theoretical concepts, p. 38-76 *in* H. Feigl and M. Scriven, eds., Minnesota Studies in the Philosophy of Science, vol. 1. The foundations of science and the concepts of psychology and psychoanalysis: Minneapolis, University of Minnesota Press, 346 p.
DUNBAR, C. O., and RODGERS, J., 1957, Principles of stratigraphy: New York, John Wiley, 356 p.
FRIEDMAN, G. M., 1961, Distinction between dune, beach, and river sands from their textural characteristics: J. Sediment. Petrology, vol. 31, pp. 514-529.
HEMPEL, C. G., 1958, The theoretician's dilemma; a study in the logic of theory construction, pp. 37-98 *in* H. Feigl, M. Scriven, and G. Maxwell, eds., Minnesota Studies in the Philosophy of Science, vol. 2. Concepts, theories, and the mind-body problem: Minneapolis, University of Minnesota Press, 553 p.
________ and OPPENHEIM, P., 1948, Studies in the logic of explanation: Phil. Sci., vol. 15, pp. 135-175.
HOOYKAAS, R., 1959, Natural law and divine miracle: Leiden, E. J. Brill, 237 p.
HORBERG, L., 1955, Radiocarbon dates and Pleistocene chronological problems in the Mississippi Valley region: J. Geol., vol. 63, pp. 278-286.
HUTTON, JAMES, 1788, Theory of the earth: Roy. Soc. Edinburgh, Tr., vol. 1, pp. 209-304.
________, 1795, Theory of the earth with proofs and illustrations, vol. I: Edinburgh, Cadell, Junior and Davies, 620 p.
KAY, M., 1951, North American geosynclines: Geol. Soc. Am., Mem. 48, 143 p.
LEET, L. D., and JUDSON, S., 1954, Physical geology: New York, Prentice-Hall, 466 p.
LEOPOLD, L. B., and LANGBEIN, W. B., 1962, The concept of entropy in landscape evolution: U. S. Geol. Survey, Prof. Paper 500-A, pp. A1-A20.
LONGWELL, C. R., and FLINT, R. F., 1955, Introduction to physical geology: New York, John Wiley, 432 p.
LYELL, CHARLES, 1881, Life, letters and journals of Sir Charles Lyell *Bart*, Mrs. Lyell, ed.: London, John Murray, 489 p.
MACKIN, J. H., 1938, The origin of Appalachian drainage—a reply: Am. J. Sci., vol. 236, pp. 27-53.
________, 1948, Concept of the graded river: Geol. Soc. Am., B., vol. 59, pp. 463-512.
MOORE, R. C., 1958, Introduction to historical geology, 2nd ed.: New York, McGraw-Hill, 656 p.
NAGEL, E., 1961, The structure of science: New York and Burlingame; Harcourt, Brace, and World, 618 p.
READ, H. H., 1957, The granite controversy: New York, Interscience publishers, 430 p.
SCRIVEN, M., 1959, Truisms as the grounds for historical explanations, pp. 443-475 *in* P. Gardiner, ed., Theories of history: Glencoe, Illinois, The Free Press, 549 p.
TURNER, F. J., 1948, Mineralogical and structural evolution of the metamorphic rocks: Geol. Soc. Am., Mem. 30, 342 p.
VON ENGELN, O. D., 1942, Geomorphology: New York, Macmillan, 655 p.
________, and CASTER, K. E., 1952, Geology: New York, McGraw-Hill, 730 p.
WELLER, J. M., 1960, Stratigraphic principles and practice: New York, Harper, 725 p.

CHAPTER FOUR

# Physical Theory and Geological Knowledge

## INTRODUCTION

More than a decade has passed since Albritton (1963) said in the introduction to *The Fabric of Geology*, "The members of the Anniversary Committee will have achieved their purpose if this collection, despite any shortcomings it may have, serves as a focal point for discussions of our role as scientists." It can now be reported that the purpose of the committee has not been achieved. Geologists were not much inclined to discuss their role as scientists before 1963 nor have they been much inclined to do so since.

Self-examination feeds upon self-doubt. In the 1960s geologists were, I think, uncertain about their role and about the level of their prestige compared with that of other natural scientists. In its impact upon science and upon the world at large, nothing had happened in geology in a century comparable with atomic physics and molecular biology. The "new tectonics" has changed this in a decade. What little philosophical reflection there was in geology has given way to an optimistic and enthusiastic pursuit of knowledge that ignores, at least for the time being, any conceptual difficulties that may arise.

But philosophical questions were raised in *The Fabric of Geology*, and they demand philosophical answers. There are, moreover, philosophical questions in the guise of geological questions that are being asked almost daily. There is a great deal of talk these days about the "revolution" in earth science. What is happening today in geology is not well understood in historical and philosophical terms. It could be better understood if some basic and, unfortunately, pervasive misconceptions about the relationship of geology to the theoretical sciences could be cleared up. It is my purpose here to examine the extent to which the theoretical sciences, principally chemistry and physics, determine the character of geological knowledge. Subsequently I shall consider the "new tectonics" in the light of this examination.

## THEORY

### *A*

This discussion of geological knowledge rests upon the assumption that geology is a historical science. Historical sciences are concerned with an account of what has happened at particular times and places in the distant and immediate past. There are different ways of getting at the past. We may simply make it up. Despite some claims to the contrary, accounts of the geologic past have never been so fanciful that it could be fairly said of them that they were made up. There have always been constraints placed upon what geologists are permitted to say about the past. Geologists go from the present to the past by means of a rational process, and for this reason geology has generally been supposed to be a legitimate part of science. This rational process consists of inferring the past from the present, or, to put it somewhat differently, constructing the past in such a way as to explain the present.

To claim that geology is a historical science is not to say that it is wholly concerned with singular descriptive statements. Most of the descriptive statements of geology refer to states and events far beyond the reach of observation. These statements are derived within inferences which must invoke general statements, be they universal, statistical, or normic. (For a discussion of these inferences, see pp. 49-56, and Kitts, 1963.)

Geologists, like other scientists, attempt to introduce systematic connections among general principles as well as among singular descriptive statements. They seek not only to explain events, but to explain the laws which are adduced to support the explanation of events. Laws are said to be explained when they are deduced from more comprehensive laws. Thus, we are told that Galileo's kinematic laws and Kepler's laws of planetary motion can be explained by deducing them from Newton's laws of mechanics and the gravitational law, together with suitable assumptions about initial and boundary conditions. If Kepler's laws and Galileo's laws are to be deduced from a set of more comprehensive laws, then the scope of these comprehensive laws must include the cases of falling bodies and planets in motion. It must be made clear somehow that falling bodies and planets in motion may be regarded as the same kinds of events. Yet the theory of mechanics does not, of course, refer explicitly to either class of events. Comprehensiveness in science is achieved by attributing to several seemingly dissimilar classes of events or objects some common hypothetical properties. Apples and planets are brought within the scope of a theory which does not in any obvious way refer to either of them, but refers to anything that can be characterized in terms of certain properties such as *mass*, designated by the theory. A distinction is often made between these widely shared, theoretical properties and more directly observable properties, and between two sets of terms employed in talking about them.

It has been generally held that the *meaning* of theoretical terms is established by their being linked, through "rules of correspondence" or "bridge principles," to observation statements. But there is no sharp distinction between what can be directly observed and what cannot, and therefore, it would seem, no sharp distinction between theoretical entities and properties on the one hand and observable entities and properties on the other. Maxwell (1962, p. 7) sums it up when he says:

> The point I am making is that there is, in principle, a continuous series beginning with looking through a vacuum and containing these as members: looking through a windowpane, looking through glasses, looking through binoculars, looking through a low-power microscope, looking through a high-power microscope, etc., in the order given. The important consequence is that, so far, we are left without criteria which would enable us to draw a nonarbitrary line between "observation" and "theory."

*B*

Geology may seem to contain an abundance of terms near the "observation" end of the continuous series. On the basis of the statement, "Wind-driven sand, like snow, tends to settle in the wind-shadow of topographic obstructions" (Dunbar and Rodgers, 1957, p. 19), we may expect certain more or less directly observable characteristics of events to be associated with one another. However, Bagnold (1941, p. 96) points out, "The movement of sand by wind is but one aspect of the wider subject of the carriage of solid particles of whatever kind by fluids in general." And we find that Bagnold in his consideration of the movement of sand grains within this wider context introduces theoretical terms, such as *viscosity*. Most of the terms which provide for the higher-level systematization in geology are not peculiar to geology, but are terms such as *viscosity*, *force*, *electron*, and *field* which are defined within the theories of other sciences.

Geologists seldom talk about deducing the principles of geology from higher-level hypotheses. They talk a great deal, however, about the relationship of geology to the "basic sciences." One sense in which those sciences are basic is that they are regarded as more comprehensive than geology. Geologists certainly regard whatever regularities they may be able to elucidate as manifestations of a deeper underlying regularity and are furthermore in substantial agreement that this deeper regularity finds its expression in physical and chemical theory. There is a widely held view that the history of a science proceeds from description of phenomena, through a state of empirical generalization, to a final state of theory formation. Whatever the virtues of this historical model for science in general, and I suspect they are very limited, it is not adequate for geology.

Geological generalizations have not been formulated outside the context of a comprehensive theory to which they are supposed to be related. In the mainstream of the geological tradition, the aim has been to formulate generalizations in such a way as not to violate the principles of physics and chemistry. These comprehensive principles serve as a guide and a restriction to geologists. The nature of the guide and restriction changes as comprehensive theory is, from time to time, reformulated.

The connection between geological statements and the theoretical concepts of physical theory is even more intimate than the above remarks would indicate. Feyerabend (1965 and elsewhere) has attacked the very legitimacy of the distinction between observation terms and theoretical terms, largely on the grounds that it presupposes that observation terms have a stable meaning independent of the theoretical context in which they occur. He says (1965, p. 180), "Words do not 'mean' something in isolation; they obtain their meanings by being part of a theoretical system." The term *geosyncline*, it must be granted, obtains its meaning from the theoretical context in which it occurs. It is not the same in Kennedy (1959) and in Holmes (1965). Feyerabend claims that even observation terms obtain their meaning from the theoretical system in which they occur. He says, for example (1965, p. 213), "According to the point of view I am advocating, the meaning of observation sentences is determined by the theories with which they are connected. Theories are meaningful independent of observations; observational statements are not meaningful unless they have been connected with theories." Consider the following geological description from Smith (1953, p. 53):

> The minimum thickness of the lower part of the Paracotos formation is estimated to be 1300 meters in the area studied. The oldest beds of this sequence are green, chloritic phyllites, composed predominantly of chlorite and quartz, with accessory zircon, zoisite, garnet, and tourmaline, and locally interbedded with gray to black, thinly bedded limestone layers. A fair degree of schistosity and lineation are evident in most of the rocks, and some show intense folding. It is believed that this is the result of nearby faulting. Cross-cutting quartz veins are locally abundant in this zone.

The terms *chlorite, quartz, limestone*, and even *rock* clearly obtain their meanings by virtue of their occurrence in a theoretical system. It would appear, on the other hand, that *minimum thickness* and *green* are relatively stable in different theoretical contexts. Be that as it may, we cannot regard geological descriptions and generalizations as being formulated in terms that have a meaning unrelated to the higher-level hypotheses to which a geologist attempts to relate them. The systematization of geological knowledge to some minimum level arises automatically out of descriptions formulated in terms already inbued with theoretical meaning.

*C*

A most significant feature of sciences with a highly developed theoretical structure is the degree to which it is possible to make logical connections among its statements. A general statement, in combination with other general statements or with singular statements, may serve as the basis for deductive derivation of other general statements, or may itself be derivable in the same way. The high regard for this systematization is reflected in Braithwaite (1953, p. 301-2), who holds that a hypothesis is to be considered lawlike only on the condition that it "either occurs in an established scientific deductive system as a higher-level hypothesis containing theoretical concepts or that it occurs in an established scientific deductive system as a deduction from higher-level hypotheses which are supported by empirical evidence which is not direct evidence for itself."

An important goal of science has been to provide theoretical structures of greater and greater comprehensiveness. This goal is sought in the name of the unity of knowledge and even in the name of the conviction, common among scientists including geologists, that the regularities encountered among observable phenomena are actually manifestations of a deeper and more fundamental regularity. To relate deductively a geologic generalization to some higher-level hypotheses is regarded as a contribution to the unity of knowledge and to the support of the conviction that there is a unity in nature itself. Important as these general considerations may be, the geologist has something of more immediate utility to gain from theoretical activity. Braithwaite suggests in the passage quoted above that we may distinguish between direct and indirect evidence for a general hypothesis. If, for example, a geological generalization could be shown to be a deductive consequence of mechanics, then empirical evidence supporting any proposition of mechanics could be counted as indirect support of the geological generalization. Many geological generalizations receive their principal support from the fact that they are, if not deducible from, at least plausible in terms of physical and chemical theory. A good deal of the empirical evidence available to support the statement, "*The immediate cause of an earthquake is the sudden break in rocks that have been distorted beyond the limit of their strength, in a process called faulting*" (Leet and Judson, 1965, p. 300), is indirect in the above sense.

With the help of a theory, we can provide indirect empirical support for a generalization referring to events whose occurrence is very rare or difficult to observe. I may claim indirect empirical support for the statement, "Rivers of ethyl alcohol are, other things being equal, less competent than rivers of water," which is to say that I may conclude on the basis of theory that events which would directly support this generalization are *possible* even though they may never occur. Observations cannot tell us what is possible and what is not. They can only tell us what is. To say that it is impossible to build a perpetual motion machine is not the same as to say that no one has ever seen one. Those who attempt to build such machines have not rejected observations, they have rejected a theory.

If geologists were to employ only their empirical generalizations in historical inference, the past would consist of events very like those in the present. By introducing theory, they can consider the degree to which it is possible for the past to differ from the present. Thornbury says (1954, p. 16), "Hutton taught that 'the present is the key to the past,' but he applied this principle somewhat too rigidly and argued that geologic processes operated throughout geologic time with the same intensity as now." I think that Thornbury is mistaken. Hutton understood clearly that our view of the past need not be rigidly restricted by the character of the present. He said, for example (1795, p. 31-32):

> It must be evident, that nothing but the most general acquaintance with the laws of acting substances, and with those of bodies changing by the powers of nature, can enable us to set about this undertaking with any reasonable prospect of success; and here the science of chemistry must be brought particularly to our aid; for this science, having for its object the changes produced upon the sensible qualities, as they are called, of bodies, by its means we may be enabled to judge of that which is possible according to the laws of nature, and of that which, in like manner, we must consider as impossible.

It is the "laws of nature" which tell us, according to Hutton, what is possible and what is impossible. Our ability to consider possibility is all that makes the pursuit of history worthwhile, for it is all that permits us to imagine that the past was, in any significant way, different from the present.

*D*

Theoretical considerations may influence geologists in a very direct and explicit way. A geologist may even begin with comprehensive laws and attempt to deduce geologically significant principles and generalizations.

The "sixth-power law" is stated by Rubey (1938, p. 121) in the following way: "The weight or volume of the largest particles that can be moved by a stream varies as the sixth power of the stream velocity." This law was first formulated, although not in this precise form, by Leslie (1823, p. 390-92).

An examination of Leslie's account of transportation reveals that his approach was almost wholly theoretical. No experiments were cited and apparently none were performed, not even as a test of the law which has been formulated on theoretical grounds. In effect Leslie asks, "Given a solid particle in the bed of a stream and given our knowledge of the behavior of physical objects as formulated in the science of mechanics, how might we expect the particle to behave?"

Rubey (1938) examined the sixth-power law, reconsidered its theoretical foundations, and tested it against experimental data provided from Gilbert (1914). He concluded (p. 134), "The analysis of Gilbert's data thus far has tended to substantiate the 'sixth-power law' for coarse sand and gravel but to indicate significant departures from the law for smaller particles." Leslie would probably not be surprised at this, for in the introduction to his previously cited work he said (pp. xii-xiii), "But the most rigorous application of mathematical reasoning will not always succeed in revealing the secrets of Nature. The philosophical inquirer must often content himself at first with seeking merely to approximate the truth." How like Bowen who said more than a century later (1928):

> It was my hope that, before anything of the kind offered was written, all of the diagrams it would be necessary to use would be determined diagrams. Yet I offer no apology for the use of deduced diagrams where this is still necessary. Vogt's pioneer work with such diagrams has more than justified their use. Attack with their aid may be regarded as a skirmishing which feels out the strength and the weakness of our adversaries the rocks, and thus lays a necessary foundation for a more serious campaign of experimental attack, concentrated upon those points where progress is most likely to be made.

For Rubey, Gilbert's experimental data serve as a test of the sixth-power law. The law is logically related to more comprehensive hypotheses, and the experimental data must relate to them also. When Rubey finds that the data depart significantly from what is expected on the basis of the sixth-power law, he does not suggest that the laws from which Leslie had derived it are false. He does not claim that the experimental data are somehow faulty. He is apparently quite willing to accept the discrepancy between law and event without suggesting that either be altered.

The relationship between laws and the comprehensive theories to which they are supposed to be related is seldom a simple one. To say that Kepler's laws of planetary motion can be deduced from Newton's laws of motion and the law of universal gravitation tends to obscure the fact that the statements that can be deduced from Newtonian theory are only approximations of Kepler's laws. As Duhem (1954, p. 193) puts it, "If Newton's theory is correct, Kepler's laws are necessarily false."

The discrepancy between what theories tell us and what we find in the "real" world is well, if often tacitly, understood. It may be accounted for in various ways, depending upon the metaphysical context in which we choose to view it. However they account for it, geologists must contend with it, and they do so most directly through experimentation. It is generally held that experiments *test* something. Geological experiments test the "fit" between theories and geologically significant configurations. They are not aimed at testing the theories which suggest the kinds of experiments to be performed.

## HISTORICAL EVENTS AND THE VERIFICATION OF THEORIES

### *A*

Theories permit us to pass beyond the limits imposed by observation to a consideration of possibility, but they clearly impose a limit upon that possibility. They prevent us from assuming that any logically conceivable event whatsoever may occur. In accepting theories we reject miracles. Historians are sometimes accused of constructing the past to fit their preconceived notions. To this extent they do just that: history, the chronicle of past events, will be constructed in such a way as not to falsify the theories to which the historian is committed. A theory is, after all, a statement to the effect that events which falsify it cannot occur. We bring to a study of the past a set of presuppositions which limits the questions we can ask about historical events. We cannot ask whether some body or sediment deposited during Cambrian time had a mass if we hold to a theory which compels us to attribute a mass to any body—past, present, or future.

To say that historians construct the past so as not to falsify certain theoretical presuppositions is not to point to a defect in historians or in their method. It is to focus upon the means historians use to find out what happened. It is to say, in a somewhat different way, what has been said before: that we infer the events of the past from the events of the present by linking them in terms of some general principles. Putting it this way calls attention to the fact that historical events play a limited role in testing theories.

Suppose, for example, a historian encounters a document containing the following passage: "On April 14, 1894, I was walking by the Times Building in New York City when I happened to glance up and see a man in a window, which was later determined to be 144 feet above the street, holding a large rock as if he were about to release it. I quickly took my stopwatch from my pocket and was able to time the fall of the rock. It fell to the street in two seconds." A historical explanation of this document might include the assumption that the law of falling bodies, which requires that a body at the earth's surface will fall 144 feet from rest in 3 seconds rather than in 2 seconds, is false. We should certainly reject such an explanation and be prepared to go to almost any lengths in order to explain the passage without resorting to a falsification of the law. We might hypothesize that the man who wrote the passage was not telling the truth, or even that someone had cleverly concealed a small rocket motor in the rock. It might be a matter of some historical interest that a man who wrote what purported to be descriptions of the events of his time was a liar, or that someone had gone to the trouble to bury a rocket motor in a rock in order to deceive some passerby. But this has no bearing on the validity of the law of falling bodies. The presupposition of the law was what led us to suspect a "trick" in the first place and thus to discover some interesting items of historical knowledge.

Suppose now that an investigation revealed that the man who had written the document which contained the description in question was well known for his veracity and that fifty reliable witnesses, all with steel tapes and stopwatches, agreed with his account of the event. We should then proceed to suggest other hypotheses to account for the description. But let us finally suppose that we were able, on the basis of wholly independent evidence, to reject every one of these hypotheses. Would we then finally reject the law of falling bodies? No, we would not. We would claim that if we knew more, or if we were more imaginative, we would be able to detect the "trick" which had made it appear that the law to which we are so deeply attached is false.

But it is not only events out of the distant historical past that we refuse to accept as falsifications of established laws. Suppose we actually witnessed a rock that fell 144 feet in 2 seconds. The suspicion that we had been tricked would occur to us just as it had in the case reported in a historical document. We would offer hypotheses in order to save the law of falling bodies, and if these hypotheses could be eliminated for lack of supporting evidence we would simply conclude that the trick had eluded us.

*B*

Can we present historical events as *instances of confirmation* for a law? We cannot if the very law we wish to test has been presupposed in inferring the event. If I am told that at some time in the past a rock was dropped from a height of 144 feet, I might then infer, by presupposing the law of falling bodies, that it reached the ground in 3 seconds. I may not then turn about and offer this event as an instance of confirmation for the very law which had been presupposed in order to infer it. Suppose on the other hand I have an account from someone who claimed to have timed the fall of the rock from 144 feet at 3 seconds. I might then offer it as an instance of confirmation for the law. This instance would, however, be regarded as offering only weak support for the law. Certainly, no law whose only instances of confirmation consisted of *events from history* would be considered well confirmed. But all instances offered to support laws and theories are accounts of *past events*. Our reluctance to accept historical events as tests for laws and theories does not stem from any mysterious quality inherent in the past. It is rather a natural outgrowth of the dictum that laws be tested under circumstances permitting the maximum possible control over, and knowledge of, the initial and boundary conditions which are, according to the laws to be tested, relevant. Because we cannot achieve such control and knowledge in the case of historical events, we are almost always in a position to accept those historical events that seem to support a theory to which we are committed and to reject those that seem to falsify it. In order to revoke the license which history grants, we have agreed that theories will be tested against events in the "present," which is to say in the proximate past.

*C*

The relevance of all of this to geology is, I trust, fairly obvious, but let us turn once again to the geologic literature. Reade (1908, p. 518) states:

> In attempts to unravel some of the weightier problems of geology it has lately been assumed that certain discordances of stratification are due to the thrusting of old rocks over those of a later geological age. Without in any

way suggesting that the geology has in any particular instance been misread, I should like to point out difficulties in accepting the explanation looked at from a dynamical point of view when applied on a scale that seems to ignore mechanical probabilities. Some of the enormous overthrusts postulated are estimated at figures approaching 100 miles. Have the authors considered that this means the movement of a solid block of rocks or rock of unknown length and thickness 100 miles over the underlying complex of newer rocks? If such a movement has ever taken place, would it not require an incalculable force to thrust the upper block over the lower, even with a clean fractured bed to move upon? Assuming that the block to be moved is the same length as the overthrust, the fractureplane would in area be 100 × 100 = 10,000 miles. I venture to think that no force applied in any of the mechanical ways known to us in Nature would move such a mass, be it ever so adjusted in thickness to the purpose, even if supplemented with a lubricant generously applied to the thrust-plane. These are thoughts that naturally occur to me, but as my mind is quite open to receive new ideas I shall be glad to know in what way the reasoning can be met by other thinkers.

A geological inference leads to the conclusion that large masses of rock have been laterally displaced over considerable distances and yet "no force applied in any of the mechanical ways known to us in Nature would move such a mass." "The mechanical ways known to us in Nature" are systematized in mechanics. A geological inference seems to require an event that the theory of mechanics forbids. We have in the words of Hubbert and Rubey (1959) the "mechanical paradox of large overthrusts." The paradox may be resolved by altering either the geological inference or the mechanical inference. In what I have characterized as the mainstream of geological thought, however, no one would hesitate over the choice. Among geologists, there is a remarkable consensus about what is more basic and fundamental and what is less so. "Basic" and "fundamental" here mean not only comprehensive but also inviolable. The geologist chooses to alter the geological inference in order to save mechanics. As soon as the decision to alter one inference to preserve the other is made, the means of achieving the alteration is suggested. In cases where the geological evidence is compelling, the resolution of the paradox will not consist of an out-of-hand rejection of the event, but rather of an alteration of the event to bring it into accord with the fundamental theory. And, in altering the event to conform to the theory, a geologist may discover something of great interest about the conditions of the past. Smoluchowski (1909, p. 204), recognizes the paradox of overthrusts and adds to it a new difficulty:

It is easy enough to calculate the force required to put a block of stone in sliding motion on a plane bed, even if its length and breadth be 100 miles, and I do not think Mr. Mellard Reade meant to use the word "incalulable" in a literal sense. . . .

Let us indicate the length, breadth, and height of the block by *a, b, c,* its weight per unit volume by *w*, the coefficient of sliding friction by *e*; then, according to well-known physical laws, a force *a, b, c, w, e* will be necessary to overcome the friction and to put the block into motion. Now, the pressure exerted by this force would be distributed over the cross-section *a, c*; hence the pressure on unit area will be equal to the weight of a column of height *b, e*. Putting $e = 0.15$ (friction of iron on iron), $b = 100$ miles, we get a height of 15 miles, while the breaking stress of granite corresponds to a height of only about 2 miles. Thus we may press the block with whatever force we like; we may eventually crush it, but we cannot succeed in moving it. The conclusion is quite striking, and so far we cannot but agree with Mr. Reade's opinion.

But are we entitled, therefore, to condemn the theory of Alpine overthrusts? I think the comparison is not quite fair. First, it may be remarked, the bed may not be horizontal but inclined; in this case the component of gravity is sufficient, at an inclination of 1:6.5, to put the block in sliding motion, and we need not apply any external pressure at all. And what seems still more important, nobody ever will explain Alpine overthrusts in any other way than as a phenomenon of rock-plasticity. Suppose a layer of plastic material, say pitch, interposed between the block and the underlying bed; or suppose the bed to be composed of such material: then the law of viscous liquid friction will come into play, instead of the friction of solids; therefore any force, however small, will succeed in moving the block. Its velocity may be small if the plasticity is small, but in geology we have plenty of time; there is no hurry.

Smoluchowski suggests, in effect, that the event may be altered to resolve the paradox by hypothesizing either that the mass of rock moved along an inclined plane or that all or some part of the block was plastic rather than rigid. These are tentative hypotheses about the character of the event. The fact that either may provide a satisfactory explanation within the confines of mechanical theory is not enough. If these hypothesized conditions are to be awarded the status of descriptions of the past, they must be supported by evidence independent of the explanation. De Sitter (1964, p. 254-55) cautions against assuming gravitational gliding in the absence of independent evidence to support it when he says: "An appropriate slope must be recognizable in the field to account for any gliding. One may of course reason, in case such a slope is absent, that the original slope has been destroyed since the gliding by the sinking of its highest portion, and if such oscillation can be proved by independent evidence, the reasoning may be sound, but in general it can hardly be accepted."

Geologists, during the past four or five decades, have invoked both gravitational sliding and reduced coefficients of friction for the rocks involved to explain overthrusting. Others have felt that the independent evidence for slopes sufficient to result in major tectonic sliding or rocks with sufficiently low coefficients of friction is far from compelling.

Hubbert and Rubey (1959, p. 162) have attempted another resolution of the paradox:

> It therefore appears that, during periods of orogeny in the geologic past, which often have affected sedimentary sections many kilometers thick, the pressure in the water contained in large parts of these sediments must have been raised to, or approaching, the limit of flotation of the overburden. This would greatly facilitate the deformation of the rocks involved, and the associated great overthrusts, whether motivated by a push from the rear or by a gravitational pull down an inclined surface, would no longer pose the enigma they have presented heretofore.

Hubbert and Rubey have, they believe, a hypothesis that would resolve the paradox once and for all. They set out to support it by finding independent evidence for high fluid pressures in rocks involved in major overthrusts (Rubey and Hubbert, 1959). Mechanics then appears to have been saved. Has the theory received additional support? In some sense. We have another class of events consistent with the theory. Once again, geologists have been led to a solution of a historical problem under the guidance of mechanics. There is the possibility, but certainly not the guarantee, that some other theory would have permitted such a solution. The force of any support for mechanics provided by Hubbert and Rubey's resolution of the paradox is weakened by the fact that the whole course of affairs leading up to their hypothesis was conditioned by the foregone conclusion that, no matter what character the events in question came to have, it would be such as to support the theory. But for the geologist, the attempt to resolve paradoxes is not directed at providing support for theories. It is directed at inferring an account of the mutable specific conditions of the past within the confines of an immutable natural order expressed in scientific theories.

*D*

We know before we begin that, no matter what the character of an event at the end of a historical inference, it must be such as to bring it into accord with certain fundamental theories or "laws of nature." There is only minor disagreement among geologists as to what constitutes the body of inviolable principles. This is not to say that geologists regard this body of general knowledge as *true*, or even subject to test. With respect to the context of historical inference, it is axiomatic.

Historical investigation is directed toward selecting from among events permitted by these fundamental theories, the events that have actually occurred. Each historical event falls into a class of events with which we are familiar by virtue of our familiarity with the theory which defines the class. It would be too much to say that we could not be surprised by some historical event that had been inferred. It would *not* be too much to say, however, that we cannot encounter a "novel event" in history—if a novel event is defined as one which violates a fundamental theory. The answer to the question, What can we discover in the past? is, Anything but an event which is novel in the above sense.

Those who have characterized geology as a "branch of natural history" may be claiming that geological inference, while it can generate a great variety of singular descriptive statements, can never transcend the theoretical apparatus it presupposes. Some geologists have claimed that this view is mistaken and that geoolgy can contribute in a rather immediate way to the development of fundamental theories. In the next section we shall examine this claim.

## HISTORICAL LAWS

### *A*

I have attempted to show that a geologist's view of the past is both expanded and limited by the theoretical context in which he operates. Theory permits him to pass beyond the limits imposed by observation, but in doing so imposes another, albeit much broader, limit upon what he may suppose to have occurred in history. Geologists will claim, however, that they have more to contribute than an addition to our knowledge of new instances of already familiar kinds of events. For example, McKelvey says (1963, p. 69): "Because it observes the results of complex natural experiments conducted on a large scale in both time and space, geology is an exploratory science that provides opportunity to observe phenomena and processes not predictable on the basis of knowledge and theory acquired and developed through the laboratory sciences."

Complexity is sometimes bemoaned as an unfortunate aspect of nature which must be tolerated. But the world *is* complex, geologists are likely to maintain. To ignore this is to ignore a fundamental aspect of reality. The events of the "real" world do not consist of frictionless systems and ideal gases. They consist of landslides and volcanic eruptions. The power of geology stems in large part from its willingness and ability to deal with complexity. Complexity vastly increases when we add to the richness of the moment the events from the "vast stretch of geologic time." Because we must presuppose some principles in order to get at past events, they *do* turn out to be new instances of already familiar kinds of events. But historical events do not simply disport themselves chaotically in the past. They are organized spatially and temporally in ways that theory has not led us to expect. Might we not discover some unfamiliar causal relationship among that complexity of familiar kinds of events?

An account of the origin of some tertiary volcanic rocks in Nevada (Riehle et al., 1972) refers to events in spatio-temporal relationship, but it provides no opportunity to discover new causal relationships. Riehle et al. (1972, p. 1392) state: "The local, lenticular, bedded, tuffaceous rocks indicate the presence of small water filled basins during the earlier eruptive phase."

Our very knowledge of these events depends upon their correlation with other events in terms of presupposed causal relationships. If a succession of events is to suggest a previously unrecognized causal relationship, the means of ordering the events in the first place must not presuppose a causal link between them.

David Hume, in his analysis of the concept of causation, claimed that the cause and effect relationship between two events is suggested by our perception of their constant conjunction. Whatever we may think of the adequacy of Hume's analysis, we can probably agree that constant conjunction is a necessary, if not sufficient, condition for the empirical support of a claim that certain kinds of events are causally related. We may determine that two events occur in conjunction by perceiving that they do. Not all conjunctions suggest causal relationship, but only those which are preserved in several instances. It would be plausible to claim that the causal association between explosive volcanic eruptions and the deposition of pyroclastic materials was suggested by the perception of their constant conjunction. But we began by asking if historical events, which cannot be perceived, might suggest in their constant conjunction hitherto unrecognized causal associations. Geologists cannot hope to formulate new causal generalizations unless they have a means of ordering events that does not presuppose causal connection among them. They must have a principle which is, in this important respect, like what might be called "Hume's principle of the conjunction of perceptions." The *law of superposition* provides them such a principle.

*B*

Stokes (1966, p. 57) has formulated the law of superposition in this way: "In any undeformed sequence of sedimentary rocks (or other surface-deposited material such as lava) each bed is younger than the one below it and older than the one above it." Geologists seem to be somewhat uncomfortable in the presence of this "law" or "principle." They recognize on the one hand that it is fundamental in historical inference and feel on the other that it is somehow trivial. Stokes goes on to say, for example: "This statement expresses such a simple and self-evident fact that it seems scarcely worth emphasizing. Nevertheless, it remains the most important generalization in the whole realm of earth history." What is simple and self-evident about the law of superposition? It is simple and self-evident that, in a set of objects that have been stacked up one after the other, any object in the stack was put down before any object above it and after any object below it. Steno did not have to discover this. It must have seemed as simple and self-evident to him as it does to Stokes. The principle is part of an array of ordinary, everyday concepts. If the principle of superposition is to be of any use, however, we must be provided not only with the abstract principle itself but also with a knowledge of the classes of things which may be considered as "objects stacked up one after the other." We cannot tell by looking at a brick wall that bricks are covered by the principle. We know this because we know the way a brick wall is built.

The geological law of superposition is an extension of the everyday law of superposition. It states, in effect, that layers of sedimentary rock may be regarded as members of a class of things that are added to a stack, one by one, from the bottom up. The justification for this hypothesis is not to be sought in its self-evident truth but in a theory of how sedimentary rocks are formed. Steno, in supporting the geological law of superposition, lays the foundations for such a theory. The geological law of superposition is a far from self-evident extension of the everyday law of superposition to a class of geologically significant entities. If it is to fulfill its critical role in historical inference, moreover, it must contain the assumption that the order of beds in an undisturbed sedimentary section reflects, not only the order in which they were added to the stack, but also the order in which they were formed. The everyday version of the law does not contain this assumption, which again must be justified by pointing to that remarkably complex and theoretical body of knowledge we call sedimentology.

Although we often speak of the "law" of superposition, we must not be misled into confusing it with a causal law of the usual sort. The law does not state that a particular bed must bear some relationship in superposition to some other bed, nor that any class of beds must bear some relationship in superposition to any other class of beds. It does not even state that sedimentary beds must be deposited in sequence. There is no implication that the character of a bed is determined by its position relative to any other bed. If there is any *causal relationship* implied by the law of superposition, it must be a relationship between bed and sequence and not between bed and bed. We could say, somewhat awkwardly, that a sequence of sedimentary beds is caused by individual beds being deposited one after the other. The law of superposition does not serve as a justification for the claim that the deposition of a bed is the cause of the deposition of the bed above it, but it does not preclude the possibility that in a given case it might be. Thus, an event inferred from a sedimentary bed can be related in space and in time to an event inferred from another sedimentary bed without supposing to begin with what they are causally related.[1]

The law of crosscutting relations is the same kind of principle as the law of superposition. It is no great discovery that an object must exist before it can be penetrated by another object. And because we have a general knowledge of firearms, we may conclude without a moment's hesitation that a block of wood containing a bullet may be regarded as an instance of the principle of crosscutting relations. Not everyone would conclude, however, that an igneous dike in a sandstone represents an instance of the class of objects crosscutting other objects. To do so requires a knowledge of geology. Like the law of superposition, the law of crosscutting relations provides a means of ordering events in time without presupposing a causal relationship between them. Physical events may also be temporally ordered without presupposing a causal relationship between them if they can be "correlated" by the use of fossils. This problem is discussed in a later chapter (see pp. 128-47).

*C*

I know of no case where a geologist has simply been able to say: "Here are two kinds of historical events whose conjunction, as inferred from the laws of superposition and crosscutting relations, is constant, and I suggest therefore that they are causally related." However, something close to this very simple situation is to be found in considerations of the association of underclays and coals. Weller (1930, p. 121) says, for example: "The intimate association of Carboniferous coals and underclays is almost universal. This association, the character of the underclay, and the not uncommon occurrence of root markings in it have been considered, for many years, evidence that the underclay represents the soil in which the coal plants grew." When geologists claim, in effect, to have discovered new causal relations among events, it is usually in much more complex situations than this.

The events alluded to in the account of Riehle et al. (1972) quoted earlier were related to one another in terms of presupposed causal generalizations, and in fact our very knowledge of these events depended upon an assumption of their causal relationship. This is not the case in the following historical account from Foose et al. (1961, p. 1164):

During the Triassic, Jurassic, and early Cretaceous, deposition within shallow seas on the stable shelf continued. Thickening of deposits during the late Cretaceous marked the beginning of more rapid shelf subsidence and initiation of early Laramide crustal instability. Part of this instability was manifested by the beginning of volcanic activity along the northeast border of the present Beartooth block and the northwestern extension of the Nye-Bowler lineament (Rouse, 1937). Livingston volcanic rocks are interbedded with rocks as old as the Upper Cretaceous Judith River formation (Parsons, 1942). Doming and possibly incipient faulting occurred at the same time along the Nye-Bowler zone and probably along much of the perimeter of the Beartooth Block, particularly the east and northeast sides. Thickening of upper Cretaceous sedimentary rocks on the south side of the Nye-Bowler zone near Red Lodge with respect to the north side indicates that that portion of the southern block moved downward during sedimentation.

We have here an account of some geological events, each of which is new to us, but all of which fall into classes of events with which we are already familiar. Although we have not before encountered *this* volcanic eruption or *this* faulting, we have encountered other events of the same kind. But there is more here than just some new instances of old familiar kinds of events. These events are ordered with respect to one another in space and time, and furthermore this ordering is not based wholly upon presuppositions of causal relationship. The assertion that major uplift of the Beartooth block occurred after the eruption of the Livingston volcanic rocks is not based upon the notion that the two events were causally related. The relationship does not patently violate any fundamental laws; on the other hand, no fundamental laws, or for that matter geological principles, would have permitted the inference that uplift of the Beartooth block would occur after the eruption of the Livingston volcanic rocks. Foose and his collaborators have discovered, in a series of historical inferences, some relationships among events that could not have been discovered in any other way. The ability of geologists to discover unexpected patterns among the events of the past is particularly significant because they claim to have discovered recurrent patterns of events which they designate by generic names.

I should like at this point to make a rather rough and ready distinction between what I shall call *primary* and *secondary historical events.* A primary historical event shall be understood to mean an event of such limited temporal extension that it might be encountered in the present or, to quantify the notion a bit, within the lifetime of a single observer. The question of whether or not such an event could occur can, in principle, be settled on the basis of observation and experiment. Secondary events, on the other hand, are so extended in time that no single observer could hope to encounter one. They are composed of primary events related in a spatial and temporal net. Some of the relationships among the primary events composing a secondary event are inferred by invoking causal laws which point to necessary relationships among events of certain kinds. But in every complex to which I shall attach the name "secondary event" there are at least two primary events related without any assumption of causal relationship between them—that is to say, related in superposition or crosscutting. Thus, in every secondary event there is some relationship that our knowledge of physical and geological theory had not led us to expect. We cannot get at these complex secondary events in the same more or less direct way that we can get at the primary events of which they are composed. If we wish to test the grounds which justify our contention that a secondary event occurred, we must test all the principles used to infer the primary events composing it and all of the principles employed in relating those primary events to each other. An answer to the question, Did this secondary event occur? or to the question, Have events of this type occurred? requires that historical inferences be performed.

There is no sharp distinction to be drawn between primary and secondary events. A rockslide is clearly a primary event and the formation of a fold mountain belt is clearly a secondary event, but what of the formation of a delta? The distinction is important enough to be worth the price of imprecision that must be paid.

When we claim that two different instances of the same kind of event are alike, we make no claim that they are alike in every aspect. We only claim that they have some characteristics in common. De Sitter (1964, p. 412) makes this clear in this characterization of a *kind* of secondary event:

> There are no two mountain chains that are characteristically alike, but many characteristic features occur repeatedly in different mountain chains.
>
> The genesis of an orthotectonic mountain chain consists of a very complex succession of events. In a complete and well-developed orogene the following phases are discernible:
>
> 1. Geosynclinal phase, which is not really of orogenic character until in its last stages; nevertheless, the initial magmatic phase belongs to its start. In its last stage the basin narrows and deepens and is characterized by flysch sedimentation.
> 2. Precursory folding phase(s), accompanied by regression of the sea.
> 3. Main folding phase, intensive compression, eventually accompanied by regional metamorphism in two stages, synkinematic and late kinematic.
> 4. Late-tectonic phases, eventually accompanied by pluton intrusion.
> 5. Posttectonic faulting phase, accompanied by volcanic activities and followed or preceded by aplanation.
> 6. Posttectonic morphogenetic phase characterized by high upheaval of aplanation surfaces.

The genesis of orthotectonic mountain chains may deviate markedly from this pattern, according to De Sitter, but he has thought it worthwhile to formulate the pattern. Why? De Sitter is not saying, "Look at these funny accidents," as one might say in pointing to two straight flushes in a single evening of poker. Presumably he, along with other geologists, believes that the relationship among the elements comprising this complex secondary event are nonaccidental.

*D*

Justification for the hypothesis that the relationships among the various elements of this immensely complex event are significant may be found in their constant or frequent association as determined from relationships in superposition and crosscutting. Might it then be legitimate to formulate historical laws which point to these significant relationships among secondary events or what may amount to the same thing, among different parts of the same secondary event? Simpson (1963, p. 29) cautions us against any such move:

> The search for historical laws is, I maintain, mistaken in principle. Laws apply, in the dictionary definition "under the same conditions," or in my amendment "to the extent that factors affecting the relationship are explicit in the law," or in common parlance "other things being equal." But in history, which is a sequence of real, individual events, other things never are equal. Historical events, whether in the history of the earth, the history of life, or recorded human history, are determined by the immanent characteristics of the universe acting on and within particular configurations, and never by either the immanent or the configurational alone. It is a law that states the relationship between the length of a pendulum—*any* pendulum—and its period. Such a law does not include the contingent circumstances, the configuration, necessary for the occurrence of a real event, say Galileo's observing the period of a particular pendulum. If laws thus exclude factors inextricably and significantly involved in real events, they cannot belong to historical science.

Much of the difficulty I find with Simpson's views of historical events and historical laws stems from my confusion about the distinction between the configurational and the immanent. He tells us (p. 28):

> Laws, as thus defined, are generalizations, but they are generalizations of a very special kind. They are complete abstractions from the individual case. They are not even concerned with what individual cases have in common, in the form of descriptive generalizations or definitions, such as that all pendulums are bodies movably suspended from a fixed point, all arkoses are sedimentary rocks containing feldspar, or all vertebrates are animals with jointed backbones.

And (Simpson, 1963, p. 28), "Laws are inherent, that is *immanent*, in the nature of things as abstracted entirely from contingent configurations, although always acting on those configurations."

It is true that the statement, "All pendulums are bodies movably suspended from a fixed point," is not a law by any usage. However, the expression, $T = 2\pi\sqrt{1/g}$, is a law by some usage—including, apparently, Simpson's. If someone were to ask, as empiricist philosophers are wont to do, What is the cash value of this statement? the answer would presumably have to include a reference to configurations. The equation is of no use whatever for empirical science unless some specification of the range of configurations to which it may be relevant is supplied. I fail to see how this could be accomplished without in some way saying what kind of a thing a pendulum is. All laws—even one so abstract as "Mass-energy is conserved"—require reference to phenomena, or to configurations, if they are to be provided with empirical support. Because a general body of common knowledge is usually assumed for those who are likely to use a law, the circumstances under which it applies are not exhaustively specified. For example, the *law of the pendulum* does not, in its usual textbook form, explicitly list all of the "contingent circumstances necessary for the occurrence of a real event." If a law does not state, or at least suggest, some limit to the configurations under which it is to apply, then we could never claim something which we do in fact claim, and that is that laws are relevant to particular "real" situations.

How are we to know that a statement to the effect that geosynclinal deposition is a necessary antecedent to the genesis of orthotectonic mountains does not express a property immanent in the material universe? "Geosyncline" and "orthotectonic mountains" must receive specification in terms of configurations, but so must "electron" and "proton." It would be silly to claim that there is no significant difference between "geosyncline" and "orthotectonic mountains" on the one hand and "electron" and "proton" on the other. However, it does seem that the difference between the two kinds of predicates is very close to the difference philosophers have tried to get at in the distinction between *observation terms* and *theoretical terms*. In *some* sense, we are likely to maintain, the concept "geosyncline" is more closely and immediately tied to configuration than the concept "proton." It is a fact, however, that any clear-cut distinction between observation terms and theoretical terms has been notoriously difficult to make. The distinction between "configurational" and "immanent" is equally difficult to make for the same reasons.

Simpson objects to the notion of historical laws on other grounds than that they cannot express immanent relationships. He says (1963, p. 29): "It is further true that historical events are unique, usually to a high degree, and hence cannot embody laws defined as recurrent, repeatable relationships." For a long time, the question of the uniqueness of historical events has been a point at issue in discussions of historical knowledge. It is certainly the case that no two events, historical or otherwise, are precisely alike in all their aspects and, in this sense, each event is unique. It is important to bear in mind, however, that, as Cohen (1942, p. 21) has put it, "The absolutely unique, that which has no element in common with anything else, is indescribable—since all description and all analysis are in terms of predicates, class concepts or repeatable relations." To emphasize this point, let us examine the description of an event from a paper by Trimble and Carr (1961) entitled, "Late Quaternary history of the Snake River in the American Falls region, Idaho." The authors state (p. 1743): "In middle or late Pleistocene time the Raft Formation was deposited, largely in a lake formed by damming of the river channel by basalt flows, possibly immediately west of the Raft River."

The description of this event is in terms of predicates and class concepts. Each property attributed to the event refers it to a class of events. It is referred to the class of depositional events, the class of events that take place largely in lakes, the class of events that take place in the vicinity of American Falls, Idaho, and the class of events that took place in Pleistocene time. There are no single properties which confer uniqueness upon an event. In assigning a number of properties to an event, we in effect assign it to a class formed by the intersection of the classes defined by each of the properties. The event in question—a member of the class of depositional events and the class of events which take place largely in lakes—is a member of the class formed by the intersection of these two, the class of depositional events which take place largely in lakes. Additional properties may place the event in classes of more limited scope. Thus, the event in our example is assigned to the class of depositional events which take place largely in lakes in the vicinity of American Falls, Idaho, in Pleistocene time. By this means, an event may ultimately be placed in a class which contains only one member. When this is accomplished, the event has been described in sufficient detail to distinguish it from all other events. But this does not remove it from any of the classes whose intersection has permitted its uniqueness to be established, or from the possibility of treatment in terms of causal generalizations that contain reference to these classes. Every property of a historical event—whether we choose to call it immanent, configurational, general, or particular—must be assigned within the context of a retrodictive inference which invokes a general, category-defining statement. Not only are we permitted to place historical events into classes, we are compelled to do so in order to describe them.

Hempel (1962, p. 18) has made a useful distinction between two senses in which the term "event" may be used. A *concrete event*, according to Hempel, is characterized by a noun, or a noun phrase, which is usually understood to refer to the overwhelming complexity of *all* the aspects of a designated event. He points out (Hempel, 1962, p. 18) that to construe the term "event" in this sense is self-defeating in scientific discourse, "for any particular event may be regarded as having infinitely many different aspects or characteristics, which cannot all be accounted by a finite set however large, of explanatory statements." I shall use the term "event" to refer to what Hempel has called a *sentential event*. A sentential event is characterized by a descriptive sentence. The properties selected for inclusion in a sentential event are those with which a scientist is able to deal, and with which he wishes to deal in a certain context. And when we claim that two events are significantly alike, we are never claiming that they are precisely alike in every imaginable aspect.

We may conceive of our knowledge of the geologic past as a knowledge of sentential events, and we may conceive of our knowledge of the geologic present in the same way. We know more of the present than of the past, but we cannot thereby claim to know the present in its concrete wholeness.

It is clear that Simpson uses the term "event" in its concrete sense. Concrete events cannot embody laws defined as recurrent, repeatable relationships; sentential events, however, can. We may define classes of sentential, secondary events and we can therefore, in principle, formulate laws which point to recurrent, repeatable relationships among them. If we were to do so, we should only claim what any scientist claims when he formulates a law linking events—that *if* certain specified kinds of conditions are realized *then* certain other specified kinds of conditions will follow. No claim is made that the particular conditions surrounding each sentential event could be exhaustively specified in a law statement, or that a law could be used to account for every aspect of any concrete event. In order to apply the law of falling bodies to two separate instances, it is not necessary to demonstrate, or to assume, that every circumstance surrounding the two instances is the same, but only to show that some circumstances are the same. Under the cover of a law, we are permitted to regard such seemingly disparate events as a man hurling himself from an airplane and a pea rolling off a knife as the same kind of event.

*F*

In an attempt to illustrate some of these points, let us turn to a classical example of an attempt to formulate a historical law in geology. Hall began it all by saying (1859, p. 69): "The line of our mountain chain, and of the ancient oceanic current which deposited these sediments, is therefore coincident and parallel; or, the line of the greatest accumulation is the line of the mountain chain." Nothing in geological or physical theory had led Hall, or anyone else, to expect the complex relationship of primary events which constitutes the secondary configuration to which he alludes. The configuration might be accidental, but Hall did not think so for he said (1859, p. 70): "I hold, therefore, that it is impossible to have any subsidence along a certain line of the earth's crust, from the accumulation of sediments, without producing the phenomena which are observed in the Appalachian and other mountain ranges." This hypothesis is lawlike in every respect. Its verification is far beyond the reach of direct test by means of observations and experiments performed in the present. It must rest upon a series of historical inferences. Do these inferences in fact provide instances of confirmation for the hypothesis? The history of the attempt to correlate subsidence of the earth's crust and accompanying sedimentation—or *geosyncline* formation, as it came to be called—with subsequent mountain formation has been fraught with misunderstanding and controversy, as Aubouin's (1965) fascinating work on the geosyncline has dramatically demonstrated. It is a fact worth noting, however, that through all this controversy a substantial number of geologists have remained convinced that the association of geosynclines and mountains is fundamentally significant. De Sitter, in the passage quoted earlier and in the one following (1964, p. 390-91), reveals his opinion that the association is no mere accident:

> I have already remarked that there is a close connection between a mountain chain and geosyncline, the geosyncline being the forerunner of a mountain chain. But this relation has proved to be only approximative. We find many deep basins which were never folded, and we also find portions of mountain chains which have never been geosynclines. The unfolded basins are numerous; the folded mountain chains not preceded by a geosynclinal phase are perhaps exceptions.

According to De Sitter, the association holds in some cases and not in others; therefore, Hall's claim of universal association is false. Geologists might have attempted, on the basis of this close but "approximative" association, to formulate a probabilistic generalization to the effect that geosyncline formation is usually, or probably, or almost always, followed by mountain formation, or even that given the occurrence of geosynclinal deposition the probability is such and such that mountain formation will follow. That no one did proceed in this way is a matter of historical record. A number of geologists did proceed by attempting to formulate the concepts "geosyncline" and "mountain" in such a way as to bring out constant or usual associations among secondary events. Let us examine the work of Kay, Aubouin, and Belousov, only three of the many geologists who have recently been involved in this effort. Kay (1951, p. 105) says of geosynclines, "Geosynclines are classified on their forms, and on the nature and sources of their contents, reflecting tectonic and volcanic environments within and beyond their borders."

Kay presents a formal classification. He attempts to provide specific criteria for the application of the term "geosyncline" and for terms referring to different kinds of geosynclines. This classification stands in dramatic denial to the notion that classification must be simply a preliminary operation preceding theory formation. Kay's categories are not mere pigeonholes. They are to be understood within the context of a theory, or at least a general hypothesis of crustal evolution. They are defined in such a way as to bring out the historical relationships among complex geological events. By defining these complex events with some minimum degree of precision, geologists hope to formulate principles or laws of association among them.

Aubouin objects to the fact that Kay and others have applied the term "geosyncline" to what he considers to be very diverse phenomena. He recommends that (Aubouin, 1965, p. 35) *"the term geosyncline should either be abandoned or its use restricted in such a way that it carries a precise meaning."* But within the context of the present discussion, this is a disagreement of minor significance. Aubouin, like Kay, is at pains to provide detailed descriptions of historical events and the geologic features which result from them, and, like Kay, proposes a classification to incorporate these descriptions. It is apparent in the following passage (1965, p. 160) that Aubouin expects that this procedure will enhance our ability to recognize significant relationships among secondary events:

> We can thus no longer be content to speak of *folding* merely as *the necessary outcome of geosynclinal evolution.* The tectonic characteristics which set geosynclinal chains apart from non-geosynclinal or intracratonal chains must be established; the basic distinctions between the different types of mountain chains will then emerge. I shall therefore endeavor to analyse the various "tectonic styles" in such a way that the unique nature of geosynclinal structures may be readily appreciated.

Belousov leaves no doubt whatever as to what he hopes to accomplish. Perhaps because of his philosophical background, he does not view the notion of historical laws with such misgivings as do many western European and American geologists. He makes his position explicit when he says (Belousov, 1962, p. 7):

> The fifth task [of geotectonics] is to discover the laws governing the succession and interrelationships of tectonic movements. For example, it has been established that folding is generally preceded by an intense subsidence of the earth's crust and the formation of a geosyncline.

Belousov does more than set the task for geotectonics. He attempts to accomplish it, as for example, in the following case (1962, p. 511):

> The third and most important regularity is that the *intensity* of the folding is determined by the thickness gradient. The greater the gradient of thickening of the deposits, the greater the intensity of the folding. The present writer considers this rule to be of great theoretical importance.

And (Belousov, 1962, p. 511-12):

> If one considers this rule and generalizes from it, one will come to the conclusion that within the geosyncline the intensive folding is due not so much to the magnitude of the subsidence and the correspondingly great accumulation of sediments as to the fact that there is an alternation of sharply differing regimes of oscillatory movements within a small area, with a corresponding transition from great thickness to small, and vice versa, characterized by a high gradient of thicknesses. By the same token, the absence of continuous folds on a platform is associated with small differences in thickness and small contrasts in oscillatory movements. This conclusion enables one to understand certain phenomena that would otherwise be obscure. It is well known that the Lower Paleozoic of the Siberian platform reaches thicknesses up to several kilometers. Such thicknesses could be taken as typical of a geosyncline. On the other hand, no geosynclinal, or continuous, folds have been observed anywhere on the platform, which differs from a geosyncline not in the thickness of its deposits, but in their nongeosynclinal distribution: the thicknesses are more or less uniform throughout, and the thickness gradient is small.

It is perfectly conceivable that the program which some geologists have undertaken in connection with the relationship between geosynclines and mountain ranges might succeed. The success would be marked by the formulation of a statement to the effect that geosyncline formation is *always* followed by mountain formation. Such a statement would, of course, presuppose an appropriate redefinition of the terms involved. The same end could be accomplished without an overt redefinition of the constituent terms if the statement were formulated normically—that is, to the effect that geosyncline formation is *always* followed by mountain formation, *except* under certain specific circumstances. Any acceptable hypothesis of geosyncline and orogeny association now would presumably have to meet the further condition of consistency with "the new tectonics."

By some criteria, such a statement might not qualify as a law; but its failure to do so could not rest on the ground that it did not point to a recurrent, repeatable relationship among events. For Simpson, it would fail, I am sure, because it referred to configurational rather than immanent properties. For others, it would fail on the ground that the stated relationship between the complex events was held to be but a consequence of the fundamental and comprehensive laws of physical science. It would thus have the logical status of an empirical generalization that could be explained in terms of physical laws. Could a generalization linking secondary historical events be invoked in an explanation or in a prediction? This would depend upon whether the user of the generalization believes that application to unexamined cases is warranted. It is clear that geologists regard the generalizations linking geosyncline formation and orogeny as more than an economical way of expressing something that could as well have been expressed in a finite conjunction of singular statements. Surely, Belousov means to claim that in all cases—past, present, and future—there is a relationship between thickness gradient and intensity of folding.

*G*

Geologists might be willing to predict on the basis of a well-formulated generalization that an orogeny would follow the formation of a geosyncline; in fact some have done so on the basis of the generalization now available. But suppose that, on the basis of the same generalization, someone claimed to have *explained* the occurrence of an orogeny by pointing to the fact that it had been preceded by the formation of a geosyncline. The parallel between historical laws and empirical generalizations can be drawn again at this point. The generalization "Ice floats in water" could be used to support the prediction "If I place this ice cube in this glass of water it will float." But suppose that I were to reply to the request for an explanation of the fact that this ice cube is floating in this glass of water by saying, "Ice floats in water and this ice cube was placed in this water." Few would consider this to be an adequate explanation. The significance of empirical generalizations does not lie in their explanatory efficacy, but rather in their relationship to laws and theories regarded as clearly having such efficacy. This is certainly true of generalizations linking secondary geological events. Geologists do not claim explanatory force for them. The search for universal relationships among secondary events is not directed toward the formulation of laws to be invoked to support inferences. In his discussion of historical laws, Simpson has been at pains to make this point. Despite my disagreements with him, I accept this important conclusion.

But this is not to say that the search for historical generalizations is unimportant. It is indeed the case that geologists are very often concerned with explaining each of the features of an event which taken together are sufficient to distinguish it from all other events. But they may also seek to explain some features of an event whose significance lies in the fact that the features are shared with other events. This is a roundabout way of saying that geologists want to explain the general regularities of history as well as the specific events of history. Historical generalizations are not so much formulated *to explain* as they are *to be explained*. A preliminary to this kind of explanation is the identification in the overwhelming complexity of the past recurring and, by implication, especially significant properties and relations. The effort to characterize historical events in such a way as to permit their being placed in theoretically significant classes, and to discover usual or constant relationships among the members of the classes, must be viewed as an effort directed toward this end. It is inconceivable that a general explanation of mountain building can be developed if each orogenic episode is considered in all of its vast uniqueness, just as it is inconceivable that a general explanation of falling bodies could have been developed if each case of a falling body had been considered in all of its historical richness.

Whatever regularities are found to obtain in the relationships among secondary events are held to be a consequence of more fundamental and pervasive regularities. If a historical generalization cannot be explained by invoking contemporary physical theory, a geologist might take the position that this theory should be altered in order to accommodate the generalization. There are two reasons why geologists never in fact take this position. First, not only geologists, but most of the rest of us, are deeply committed to the "fundamental laws of nature" contained in the physical theory of our time. These laws are not overthrown in the course of the day-to-day practice of historical science. They are overthrown in episodes so rare and dramatic that we have chosen to call them "revolutions." Second, although the inference of some of the aspects of secondary events and of their relationship to one another *does not* presuppose the laws of physical theory in any immediate way, the inference of many other aspects of these events *does* presuppose these laws. They are not novel with respect to physical theory because, among other things, the primary events of which they are composed and many of the relationships among them presuppose physical theory in a direct way. Historical generalizations are by no means structured in a way that is neutral with respect to any conceivable higher level hypotheses which could be invoked to explain them. They have been formulated in concepts theoretically relevant to a group of specific theories.

The geologist, because he has in effect decided that he will not formulate new theoretical propositions in his attempt to explain historical generalizations, proceeds much as he does in his attempt to explain particular events, by suggesting sets of boundary and initial conditions which, together with the presumption of physical laws, will serve to explain his historical generalizations. In so doing he will discover something of historical interest. Part of what we know about the crust and upper mantle is based upon a requirement that they be of such a character as to permit geosynclines to occur, to permit mountains to form, and to permit mountains and geosynclines to be associated in a nonaccidental way.

## CONCLUSION

In order to place some restriction upon what may be supposed to have occurred in the past and thereby to avoid chaos in history, a geologist must hold theoretical presuppositions. It is not logically or empirically necessary that geologists operate within the conceptual framework of the contemporary physical theory. The theory is not imposed upon geologists; they impose it upon themselves. There are, of course, compelling historical and philosophical reasons for choosing this conceptual background rather than some other. Throughout the history of geology, physical theory has provided an immediately available inferential apparatus of great flexibility and demonstrated utility. But this has probably not been the most compelling reason for the geologist's decision to proceed under the umbrella of the fundamental theories of his time. Most geologists have not regarded these theories as mere optional formulations. The principles of chemistry and physics and biology are more than useful "inference tickets." They are held to be "true" or "nearly true" statements about reality.

The unproblematic background provided by contemporary theory imposes a restriction upon the statements which may be admitted to the body of geological knowledge. Geologists have objected to this view because it seems to relegate geology to a position of dependence upon and, by implication, inferiority to other sciences. It is, after all, a venerable tradition in our culture that principles, laws, theories, universals, and essences are held in the highest esteem and that those who formulate them are accorded a favored place in our history. According to the same tradition, those who deal with particulars, which is to say with history, are engaged at a lower level of intellectual activity, albeit a sometimes interesting and useful one. The science of geology rests firmly and necessarily upon a repudiation of this tradition. For the geologist, historical knowledge is the goal and theory is the means of achieving it.

Geologists are not alone in operating within the confines of an elaborate set of general presuppositions. The scientific community is not anything like equally divided between those who unquestioningly accept the theory of their day and those who set out to alter that theory. It is composed of a vast number of physicists, chemists, biologists, astronomers, and geologists who operate against an unproblematic background and a scant handful of scientists who are in any immediate way involved in the task of altering fundamental theoretical structure. And yet when we enumerate the scientists who did alter theories, they turn out to be associated not with geology but with other scientific disciplines. Newton was a physicist. But he stands in much the same conceptual relationship to geologists as he does to physicists. Newton formulated a theory to which almost all scientists subscribed for over two hundred years. What he did was not much more like ordinary physics than it was like geology.

It is tempting to simply say that geology is, in Kuhn's (1962) sense, "normal science." Kuhn's "normal science" is, if I understand him correctly, *directed at* strengthening, or elaborating, or articulating the paradigm upon which it is based. Geology presupposes a paradigm, but it is not *directed at* the paradigm at all. Geological research may strengthen or elaborate or articulate a paradigm, but it is not undertaken in order to do so. (For a discussion of the relevance of Kuhn's concept of scientific revolution to contemporary geology see pp. 115-27.)

No geologist need harbor the view that the generation of hypotheses and descriptions of events against an unproblematic background of theory is a rote process requiring no creativity. Theories are ambiguous, and the world of experience is complex. For this reason, nothing of historical interest will be discovered as a rigorous deductive consequence of theory. Significant items of geological knowledge, like significant mathematical theorems, arc not *discovered* deductively although an effort may be made to *justify* them deductively. Theory may serve to guide us, but only the practice of historical science can hope to approach an account of the variety and richness of the world.

REFERENCES CITED

ALBRITTON, CLAUDE C., JR., 1963, Preface, *in* ALBRITTON, CLAUDE C., JR., ed., The fabric of geology: Reading, Mass., Addison-Wesley, p. v-vii.
AUBOUIN, J., 1965, Geosynclines: Amsterdam, Elsevier, 335 p.

BAGNOLD, R. A., 1941, The physics of blown sand and desert dunes: London, Methuen, 265 p.

BELOUSOV, V. V., 1962, Basic problems in geotectonics: New York, McGraw-Hill, 816 p.

BOWEN, N. L., 1928, The evolution of igneous rocks: Princeton, N.J., Princeton Univ. Press, 332 p.

BRAITHWAITE, R. B., 1953, Scientific explanation: Cambridge, Cambridge Univ. Press, 316 p.

COHEN, M. R., 1942, Causation and its application to history: Jour. History Ideas, v. 3, p. 12-29.

DUHEM, P., 1954, The aim and structure of physical theory: Princeton, N.J., Princeton Univ. Press, 344 p.

DUNBAR, C. O., and RODGERS, J., 1957, Principles of stratigraphy: New York, Wiley, 356 p.

FEYERABEND, P. K., 1965, Problems of empiricism, *in* COLODNY, R. G., ed., Beyond the edge of certainty: Englewood Cliffs, N.J., Prentice-Hall, p. 145-260.

FOOSE, R. M.; WISE, D. V.; and GARBARINI, G. S., 1961, Structural geology of the Beartooth Mountains of Montana and Wyoming: Geol. Soc. America Bull., v. 72, p. 1143-1172.

GILBERT, G. K., 1914, The transportation of debris by running water: U.S. Geol. Survey Prof. Paper 86, 263 p.

HALL, J., 1859, Descriptions and figures of the organic remains of the lower Helderberg Group and the Oriskany Sandstone, *in* Natural history of New York, Paleontology, v. 3, New York Geol. Survey, p. 1-532.

HEMPEL, C. G., 1962, Explanation in science and in history, *in* COLODNY, R. G., ed., Frontiers of science and philosophy: Pittsburgh, Univ. Pittsburgh Press, p. 7-33.

HOLMES, A., 1965, Principles of physical geology: New York, Ronald, 1288 p.

HUBBERT, M. K., and RUBEY, W. W., 1959, Role of fluid pressure in mechanics of overthrust faulting. I. Mechanics of fluid-filled porous solids and its application to overthrust faulting: Geol. Soc. America Bull., v. 70, p. 115-166.

HUTTON, J., 1795, Theory of the earth with proofs and illustrations: Edinburgh, William Creech, 2 v., 620 + 567 p.

KAY, M., 1951, North American geosynclines: Geol. Soc. America Mem. 48, 143 p.

KENNEDY, G. C., 1959, The origin of continents, mountain ranges, and ocean basins: Am. Scientist, v. 47, p. 491-504.

KITTS, D. B., 1963, The theory of geology, *in* ALBRITTON, C. C., ed., The fabric of geology: Reading, Mass., Addison-Wesley, p. 49-68.

KUHN, T. S., 1962, The structure of scientific revolutions: Chicago, Univ. Chicago Press, 172 p.

LEET, L. D. and JUDSON, S., 1965, Physical geology: (3rd ed.); Englewood Cliffs, N.J., Prentice Hall, 406 p.

LESLIE, J., 1823, Elements of natural philosophy: Edinburgh, W. & C. Tait, 406 p.

MCKELVEY, V. E., 1963, Geology as the study of complex natural experiments, *in* ALBRITTON, C. C., ed., The fabric of geology: Reading, Mass., Addison-Wesley, p. 69-74.

MAXWELL, G., 1962, The ontological status of theoretical entities, *in* FEIGL, H., and MAXWELL, G., eds., Minnesota studies in the philosophy of science: Minneapolis, Univ. Minnesota Press, v. 3, p. 3-27.

READE, J. M., 1908, The mechanics of overthrusts: Geol. Mag. new ser., dec. 5, v. 5, p. 518.

RIEHLE, J. R.; MCKEE, E. H.; and SPEED, R. C. 1972, Tertiary volcanic center, west-central Nevada: Geol. Soc. America Bull., v. 83, p. 1383-1396.

RUBEY, W. W., 1938, The force required to move particles on a stream bed: U.S. Geol. Survey Prof. Paper 189-E, p. 120-140.

———, and HUBBERT, M. K., 1959, Role of fluid pressure in mechanics of overthrust faulting. II. Overthrust belt in geosynclinal area of western Wyoming in light of fluid pressure hypothesis: Geol. Soc. America Bull., v. 70, p. 167-205.

SIMPSON, G. G., 1963, Historical science, *in* ALBRITTON, C. C., ed., The fabric of geology: Reading, Mass., Addison-Wesley, p. 24-48.

SITTER, L. V. DE, 1964, Structural geology (2d ed.): New York, McGraw-Hill, 551 p.

SMITH, R. J., 1953, Geology of Los Teques-Cua region, Venezuela: Geol. Soc. America Bull., v. 64, p. 41-64.

SMOLUCHOWSKI, M. S., 1909, Some remarks on the mechanics of overthrusts: Geol. Mag., new ser., dec. 5, v. 6, p. 204-205.

STOKES, W. L., 1966, Essentials of earth history (2d ed.): Englewood Cliffs, N.J., Prentice-Hall, 468 p.

THORNBURY, W. D., 1954, Principles of geomorphology: New York, Wiley, 618 p.

TRIMBLE, D. E., and CARR, W. J., 1961, Late Quarternary history of the Snake River in the American Falls region, Idaho: Geol. Soc. America Bull., v. 72, p. 1739-1748.

WELLER, J. M., 1930, Cyclical sedimentation of the Pennsylvanian period and its significance: Jour. Geology, v. 38, p. 97-135.

CHAPTER FIVE

# Grove Karl Gilbert and the Concept of "Hypothesis" in Late Nineteenth-Century Geology

## *I*

In the preface of their widely used textbook of elementary geology, Leet and Judson (1954) say, "Originally, geology was essentially descriptive, a branch of natural history. But by the middle of the twentieth century it had developed into a full-fledged physical science making liberal use of chemistry, physics, and mathematics, and in turn contributing to their growth." This kind of statement is not uncommon in the twentieth century, nor was it uncommon in the nineteenth. Geologists have been remarkably apologetic about their discipline, although they have often expressed a hope, and sometimes even a promise, that things would get better. Many outside geology have expressed a similar view. The historians Basalla, Coleman, and Kargon entitle the section on geology in their *Victorian Science* (1970) "Geology Becomes a Science," clearly implying that before the late nineteenth century it had been something else. It is far from clear what is being got at in these characterizations of geology, but a key to it all seems to lie in the term "descriptive." To describe something, according to my dictionary, is to represent it in words. There is apparently no restriction as to the subject of the representation. But usually, in science at least, we take the subject of a description to be a particular object or event rather than a universal condition. Thus it is quite natural to say that we describe a sunset and at least a little strange to say that we describe the laws of physics. When someone says that geology is descriptive, he usually means to convey the idea that the body of geologic knowledge consists largely or wholly of statements about particular objects and events. An examination of the geologic literature of the nineteenth and twentieth centuries will reveal that geology is indeed overwhelmingly descriptive in this sense. I do not know why anyone should deny it or, having recognized it, should apologize for it. Geologists have always made a great deal of the fact that geology is historical. To say that geology is descriptive in the above sense is just another way of saying that it is historical. The historical concern of geologists is revealed in their pre-occupation with spatio-temporal location or, to use a convenient term introduced by Simpson (1963), with configurations. Geologists, among all scientists, have elevated particulars to the status of significant items of knowledge, and in doing so have departed from an earlier tradition which held that the events which embody history are ephemeral, contingent, and even unreal.

But there is an extreme version of the view that geology is descriptive. It is the view that geology is *merely* descriptive. It holds that geology consists largely, or wholly, of reports of direct observation. This is so patently absurd that one is bound to ask how anyone could believe it. I feel almost apologetic in noting at this point that accounts of the past and accounts of the present are formulated in the same descriptive terms. The descriptive vocabulary of geology does not, by itself, call attention to the inferential gap between assertions about the past and assertions about the present. This obvious feature of historical discourse has led geologists to an unjustified sense of familiarity with the past and has made it possible for some, both in and out of geology, to ignore the critical distinction between descriptions of the present and descriptions of the past. To focus upon this distinction is to focus upon the central methodological problem of geology. How do we get from the present to the past, or, more generally, how do we derive and test singular descriptive statements?

There are different ways of getting at the past. We may, for example, simply make it up. Despite some claims to the contrary, accounts of the geologic past have never been so fanciful that it could be fairly said that they were made up. There have always been constraints placed upon what geologists are permitted to say about the past. They go from the present to the past by some rational process, and it is for this reason that geology has been generally supposed to be a legitimate part of science.

## *II*

The late nineteenth century was a period of inactivity in the philosophy and methodology of geology and so contrasted sharply with the first half of the century, which had seen an enormous amount of theological, philosophical, and methodological discussion concerning geology. Perhaps philosophical interest in historical biology after 1860 had detracted from what by then might have seemed to be the less problematic issue of historical geology. Another factor, more important in North America than in Europe, was that geology had entered an intensely exploratory phase. As the exploration of the West proceeded, a vast area of well-exposed rock became available for the careful examination of geologists. Geology became an activity in which running rivers and climbing mountains were at least as important as philosophical reflection. Coincident with this exploratory phase and, I think, related to it was a deep commitment to field work and meticulous observation as the only way to geological knowledge.

I offer no apology for choosing to discuss this period of poverty in geological philosophy and methodology. The early nineteenth century has received a great deal of attention from both historians and geologists, but the last half of the century has been little studied. My primary purpose in focusing upon this period is not, however, simply to fill a gap in our knowledge of the history of geology. Rather, I want to examine the establishment, during the last quarter of the nineteenth century, of a methodological view that persists to this day.

I have chosen to discuss Grove Karl Gilbert, who addressed himself to issues connected with the central methodological problem of geology. Gilbert was born in Rochester, New York, in 1843 and graduated from the University of Rochester in 1862. He taught school for less than a year and then accepted a job as assistant in Ward's Cosmos Hall, which later became Ward's Natural Science Establishment. On the basis of the limited experience in geology acquired at Ward's, he was taken on as an assistant with J. S. Newberry's survey of Ohio in 1870. He spent 1871, 1872, and 1873 with the Wheeler Survey, and in 1876 he joined the Powell Survey. He continued to work under Powell and King until 1879, when the United States Geological Survey was established. He joined the Survey and quickly rose to the rank of chief geologist, a position which he held until his death in 1918. If nothing else, this brief biographical sketch will serve to show that Gilbert was close to the heart of geological activity in the United States for nearly fifty years. He made significant contributions on a variety of subjects including geomorphology, structural geology, and economic geology.

Gilbert's principal methodological work, entitled "The Inculcation of Scientific Method by Example," was presented as the presidential address before the American Society of Naturalists in December, 1885, and was published in the *American Journal of Science* the following year (1886). Gilbert regarded it as an account of science as a whole, not of geology alone. His view of science is most explicitly stated in the following passage (p. 286):

> It is the province of research to discover the antecedents of phenomena.

This is done with the aid of hypothesis. A phenomenon having been observed, or a group of phenomena having been established by empiric classification, the investigator invents an hypothesis in explanation. He then devises and applies a test of the validity of the hypothesis. If it does not stand the test he discards it and invents a new one. If it survives the test, he proceeds at once to devise a second test, and he thus continues until he finds an hypothesis that remains unscathed after all the tests his imagination can suggest.

There is nothing particularly novel about Gilbert's view of science. At least it may not seem so at first glance. It differs, however, from the views held by physicists who were Gilbert's contemporaries. Despite substantial philosophical differences among late nineteenth-century physicists concerning the nature of science, most of them appear to have agreed that the central issues of methodology revolved about the concepts *law* and *theory*. Gilbert does not mention these concepts in the quoted passage. Perhaps there is covert reference to them, in the terms *hypothesis* and *antecedent*. Gilbert's use of these terms elsewhere, however, reveals that he uses them only to refer to particulars. "If the hypothetical antecedent is a familiar phenomenon," he says (p. 287), "we compare its known or deduced consequences with A, and observe whether they agree or differ." Gilbert's antecedents are not only logical antecedents; they are temporal antecedents. The term *hypothesis* also stands for initial conditions rather than for laws or principles. "Take first the hypothesis that the crust of the earth, floating on a molten nucleus, rose up in the region of the basin when its weight was locally diminished by the removal of the water of the lake"[1] (p. 296). So far as I have been able to determine, the use of the term *hypothesis* to stand for particular antecedent conditions is virtually universal among nineteenth- and twentieth-century geologists.

We must be careful not to make too much of this point. In a very important respect, geological science is a search for the temporal antecedent of phenomena. Antecedents are discovered with the aid of hypotheses, which consist, for Gilbert, of conjectures about specific initial conditions. Gilbert's account of science may simply reflect the geologist's preoccupation with the derivation and testing of singular descriptive statements. Most of us hold, of course, that the derivation and testing of singular descriptive statements require the mediation of statements of general form, and we might suppose that Gilbert held this also. Let us examine his account more closely for evidence that beneath his primary concern for description there lies the recognition of a general or theoretical apparatus.

The term *law,* which figures so importantly in discussions of the methodology of science of Gilbert's time, may suggest reference to general or theoretical notions (p. 296).

> It is known [Gilbert states] that the density of the earth's material increases downward, for the mean density of the earth, expressed in terms of the density of water, is about 5.5, while that of the upper portion of the crust is about 2.7. Nothing is known however of the law under which the density increases, and nothing is known as to the depth of the zone at which matter is sufficiently mobile to be moved beneath the Bonneville basin.

Or take Gilbert's use of the term *postulate* (p. 291):

> The verdict of the barometer was that the southerly shoreline was somewhat higher than the northerly, but the computations necessary to deduce it were not made until the mutual continuity of the two shorelines had been ascertained by direct observation. The barometric measurement was therefore superseded as an answer to the original question, but it answered another which had not been asked, for it indicated that the ancient shore at one point had come to stand higher than that of another. The postulate of horizontality was thus overthrown.

As I have already remarked, the terms *hypothesis* and *antecedent* are sometimes used to designate specific initial conditions. The terms *law* and *postulate* almost never are. *Induction* may be used to designate an argument which leads to a general conclusion, or one which leads to a particular conclusion. In Gilbert's day, it was almost always used in connection with the "discovery" of laws and theories. In contrast, Gilbert used it in a wholly particular sense (p. 299):

> The condition of the interior of the earth is one of the great problems of our generation. Those who have approached it from the geologic side have based a broad induction on the structural phenomena of the visible portion of the earth's crust, and have reached the conclusion that the nucleus is mobile.

This "broad induction" begins with particulars: "the structural phenomena of the visible portion of the earth's crust." But it does not lead to a general conclusion. It leads instead to a singular descriptive statement about the earth's interior. Consider another instance of Gilbert's use of "induction" (pp. 291-92):

> It is one of the great inductions of geology that as the ages roll by the surface of the earth rises and falls in a way that may be called undulatory. I do not refer to the anticlinal and synclinal flexures of strata, so conspicuous in some mountainous regions, but to broader and far greater flexures which are inconstant in position from period to period. By such undulations the

Tertiary lake basins of the Far West were not only formed but were remodeled and rearranged many times. By such undulations the basin of Great Salt Lake was created.

The conclusion that "the earth's crust rises and falls in a way that may be called undulatory" could be given a general interpretation. It does not, however, figure as a general term in Gilbert's argument. Its most precise formulation would consist of a conjunction of singular descriptive statements, which is to say, a detailed historical account.[2]

However significant the above passages may be, in none of them does Gilbert directly address himself to the central methodological problem of geology. In one place he comes close to explicitly considering it (p. 287):

> Given a phenomenon, A, whose antecedent we seek. First we ransack the memory for some different phenomenon, B, which has one or more features in common with A, and whose antecedent we know. Then we pass by analogy from the antecedent of B, to the hypothetical antecedent of A, solving the analogical proportion: as B is to A, so is the antecedent of B to the antecedent of A.

I do not propose to discuss the logic of analogic arguments. I only wish to point out that Gilbert sees argument by analogy as a means of justifying a move from event to event by invoking a particular event rather than a principle or law.

*III*

Gilbert was not a philosopher, nor was he, like Hutton, a theoretical geologist of first rank. It is hardly surprising that he fails to give a complete account of geological inference. Theoretical statements do play a role in geology, and any satisfactory discussion of geological method must take them into account. Apparently Gilbert realizes this critical role, although he nowhere states it. He uses the term *deduction*, and he must understand that deductive arguments rest upon a premise of universal form. It is remarkable, however, that nowhere does Gilbert explicitly provide for a general foundation. Furthermore, by using terms for the singular and particular that stand for the general and theoretical in other contexts, he effectively cuts himself off from an analytical apparatus that could permit him to do so.

Gilbert's philosophical antecedents are obscure. Is he an inductivist or a deductivist? It is almost absurd to ask. He seems to be operating outside any familiar philosophical tradition, and his frequent use of terms with a more or less established meaning within these familiar traditions does not conceal the fact. Nevertheless there are compelling reasons to take Gilbert seriously. Almost unanimously, contemporary geologists agree that he had something important to say, and many of them hold that he succeeded in getting at the critical problems of geological knowledge. *The Fabric of Geology*, published in 1963 to commemorate the seventy-fifth anniversary of the Geological Society of America, is filled with allusions to Gilbert as the premier methodologist of modern geology. The least that we can hope to get from a study of Gilbert's notions about science is an insight into the methodological views of three generations of geologists. It is interesting to note that many of the geologists who extol the virtues of Gilbert profess to find nothing in contemporary philosophy of science which illuminates the problems of geological knowledge. How can we account for the strong appeal of Gilbert's methodology of geology where after all it counts the most, and that is among practicing geologists?

## *IV*

Describing a prevailing attitude in the early eighteenth century, Lovejoy (1960) says, "The process of time brings no enrichment of the world's diversity; in a world which is the manifestation of eternal rationality, it could not conceivably do so. Yet it was in precisely the period when this implication of the old conceptions became most apparent that there began a reaction against it." Geology was intimately involved in this reaction. Geology is the study of change and may, for some, become a study of the "enrichment of the world's diversity."

The reaction of which Lovejoy speaks raised grave philosophical questions which entailed serious methodological problems. If the process of time brings an enrichment of the world's diversity, then what limit is to be imposed upon that diversity? Some limit must be imposed. To impose no limit at all is to remove any constraint upon what may be supposed to have occurred in the past and, in effect, to remove the writing of geologic history from science, and indeed from rational discourse. For the geologist, this is no issue for idle philosophical speculation. He encounters it every time he performs a historical inference.

The classical debates of late eighteenth- and early nineteenth-century geology concern the character and magnitude of the restriction imposed by the immutable natural order upon the events of history. The solution presented by the two towering figures of this period, Hutton and Lyell, is a conservative one. To the question, "To what extent does the natural order permit an enrichment of the world's diversity in time?" their answer is, in effect, "To no great extent." In a recent article on Lyell's *Principles*, Rudwick states (1970, pp. 32-33):

Yet all these areas of emphasis are used in the strategy of the ***Principles,*** not as arguments of interest in themselves, but as tactical devices to be deployed in the service of a uniformitarian system of earth history.

It is surely in Lyell's commitment to this system that we should look for a key to his reluctance to accept a progressionist view of the history of life or, ultimately, a transmutationist view of the mechanism of that progression. For it is ironic that the uniformity that later generations of geologists and biologists came to accept from Lyell was that of this actualistic methodology; they came to reject the uniformity of his steady-state system in favor of a development system much closer to that of his directionalist opponents.

The directionalist view began to pervade geology in the late nineteenth century, and Gilbert's methodology should be viewed with this in mind. The major philosophical problem for the directionalist is the accommodation of real historical change, an increase in the world's diversity, with a real immutable order. Darwin provided a solution to the problem for the biological directionalists. The principles of Darwinian theory permit, indeed require, a highly directional account of history. Geologists had achieved no such dramatic solution to a difficult problem. It is not clear that they even saw it as a problem. Gilbert apparently did not. But there is a suggestion in the work of Gilbert, and in the work of other geologists in his time and later, that they felt that a rigid theoretical structure posed a threat to progressive history. They seem to have thought that an ultimate historical solution might be found in a method that permitted the justification of inferences from event to event without the invocation of any general apparatus whatever. This judgment suffers at least two defects as an immediate explanation of Gilbert's particularization of method. First, there is nothing in Gilbert's work to indicate that he was consciously committed to directionalism or that he was aware of the increasing tendency of geologists to give directional accounts of history. More important is the fact that geologists do routinely invoke generalizations and theories despite the reluctance of geological methodologists to discuss the procedure, and geologists, including Gilbert, reveal that they are perfectly aware of it in their day-to-day work. To accept the view that Gilbert was attempting to enhance directionalist geology by finding a method that circumvented theory is to suppose that Gilbert simply overlooked the fact that he and his fellow geologists employed laws and theories as instruments of historical inference.

Geological observations and principles are formulated wholly within the context of a complex system of general preconceptions, so complex a system that one could not hope to identify all of its components with any reasonable effort. There is, however, a readily identifiable part of this system. It is what geologists consider to be the most fundamental and comprehensive principles contained in the physical theory of their day. These principles are regarded as wholly unproblematic for the purposes of geological inference.

It is neither logically nor empirically necessary that geologists operate within the conceptual framework of contemporary physical theory. The theory is not imposed upon geologists; geologists impose it upon themselves. Throughout the history of geology, physical theory has provided an immediately available inferential apparatus of great power and demonstrated utility. But this is not the primary reason for the geologists' decision to proceed under the umbrella of the fundamental theory of their time. Geologists do not regard these theories as optional formulations. The laws of chemistry and physics are more than "inference tickets." For most geologists they are "true" or "nearly true" statements about reality. In the mainstream of geology, the scientific character of the discipline is regarded as guaranteed by its demonstrable connection with physical theory. Physical theory is applied in geological inferences so directly and so obviously that it simply cannot be overlooked. Let us consider an example from Reade's discussion of overthrusts (1908, p. 518):

> In attempts to unravel some of the weightier problems of geology it has lately been assumed that certain discordances of stratification are due to the thrusting of old rocks over those of a later geological age. Without in any way suggesting that the geology has in any particular instance been misread, I should like to point out the difficulties in accepting the explanation looked at from a dynamical point of view when applied on a scale that seems to ignore mechanical probabilities. Some of the enormous overthrusts postulated are estimated at figures approaching 100 miles. Have the authors considered that this means the movement of a solid block of rock or rocks of unknown length and thickness 100 miles over the underlying complex of newer rocks? If such a movement has ever taken place, would it not require an incalculable force to thrust the upper block over the lower, even with a clean fractured bed to move upon? Assuming that the block to be moved is the same length as the overthrust, the fracture-place would in area be $100 \times 100 = 10{,}000$ miles. I venture to think that no force applied in any of the mechanical ways known to us in Nature would move such a mass, be it ever so adjusted in thickness to the purpose, even if supplemented with a lubricant generously applied to the thrust-plane. These are thoughts that naturally occur to me, but as my mind is quite open to receive new ideas I shall be glad to know in what way the reasoning can be met by other thinkers.

A geological inference leads to the conclusion that large masses of rocks have been laterally displaced over considerable distances and yet "no force applied in any of the mechanical ways known to us in Nature would move such a mass." "The mechanical ways known to us in Nature" are systematized in mechanics. A geological inference seems to require an event that the theory of mechanics forbids, and we have in the words of Hubbert and Rubey (1959) the "mechanical paradox of large overthrusts." The paradox may be resolved by altering either the geological inference or the mechanical inference. In the mainstream of geological thought, no one hesitates over the choice. There is a remarkable consensus about what is more basic and fundamental and what is less so. "Basic" and "fundamental" here mean not only comprehensive but also inviolable. The geologist chooses to alter the geological inference in order to save mechanics. As soon as the decision to alter one inference to preserve the other is made, the means of achieving the alteration is suggested. In cases where the geological evidence is compelling, this resolution of the paradox will not consist of an out of hand rejection of the event, but rather of its alteration to bring it into accord with the fundamental theory. And in altering the event to conform to the theory, a geologist may discover something of great interest about the conditions of the past. Smoluchowski recognizes the paradox of overthrusts and adds a new dimension to it (1909, pp. 204-5):

> It is easy enough to calculate the force required to put a block of stone in sliding motion on a plane bed, even if its length and breadth be 100 miles, and I do not think Mr. Mellard Reade meant to use the word "incalculable" in a literal sense. . . . Let us indicate the length, breadth, and height of the block by *a, b, c,* its weight per unit volume by *w,* the coefficient of sliding friction by *e;* then, according to well-known physical laws, a force *abcwe* will be necessary to overcome the friction and to put the block into motion. Now, the pressure exerted by this force would be distributed over the cross-section *ac;* hence the pressure on the unit area will be equal to the weight of a column of height *be*. Putting $e = 0.15$ (friction of iron on iron), $b = 100$ miles, we get a height of 15 miles, while the breaking stress of granite corresponds to a height of only about 2 miles. Thus we may press the block with whatever force we like; we may eventually crush it, but we cannot succeed in moving it. The conclusion is quite striking, and so far we cannot but agree with Mr. Reade's opinion.

Smoluchowski goes on to suggest that the paradox may be resolved by hypothesizing either that the mass of rock moved along an inclined plane or that all or some part of the block was plastic rather than rigid. During the past four or five decades, geologists have invoked both gravitational sliding and reduced coefficients of friction for rocks involved to explain overthrusting. Others have felt that the independent evidence for slopes sufficient to result in major tectonic sliding of rocks with sufficiently low coefficients of friction is far from compelling. Hubbert and Rubey (1959, p. 162) attempted another resolution of the paradox:

> It therefore appears that, during periods of orogeny in the geologic past, which often have affected sedimentary sections many kilometers thick, the pressure in the water contained in large parts of these sediments must have been raised to, or approaching the limit of flotation of the overburden. This would greatly facilitate the deformation of the rocks involved, and the associated great overthrusts, whether motivated by a push from the rear or by a gravitational pull down an inclined surface, would no longer pose the enigma they have presented heretofore.

Hubbert and Rubey believe they have a hypothesis that will resolve the paradox once and for all. They set out to support it by finding independent evidence for high fluid pressures in rocks involved in major overthrusts.

The geologists who attempted to resolve the paradox of major overthrusts were directed at every juncture by their theoretical preconceptions. Their observations and descriptions and inferences were formulated in terms already imbued with theoretical significance. This is not to suggest that what they did was trivial or insignificant. They presented hypotheses and tested them, and in the process made significant contributions to geologic knowledge. But through it all there was one pervading hypothesis that was not subject to test at all. Mechanics was not tested against geological events. Geological events were tested against mechanics.

It is clear then that the theories of physics and chemistry are explicitly, obviously, and directly applied to problems in geology. How then can a geologist find Gilbert's account of geological method, which takes no account whatever of the role of theory, so appealing? The answer lies, I think, in the fact that geologists tend to see two distinct levels, or phases, of historical inference. Gilbert's method seems to account very well for the first phase, which consists of the initial step from the present to the past. If a geologist could be induced to admit that this step must be justified by laws and generalizations, he would be likely to claim that they are "self-evident" or "trivial" or even that they consist of "truisms." In the second phase of historical investigation, the events of the geologic past are "explained," or "interpreted," or "understood" in terms of physical theory.

Let us briefly reconsider the case of major alpine overthrusts. The inference which led to the conclusion that blocks of the earth's crust have been displaced laterally belongs to the first phase of historical investigation. A critical examination reveals that a number of principles must be adducted to justify this inference; to mention one, *the law of superposition.* This law asserts that in an undisturbed sequence of sedimentary rocks each bed is younger than the one below it and older than the one above it. For centuries geologists have been telling their students that the law of superposition is self-evident and have thereby done Steno, who formulated the law, and themselves, who use it every day, a great injustice. It is self-evident, I suppose, that when objects are stacked up one after the other, the objects lower in the stack were put down earlier. It is not self-evident, however, that sedimentary rocks may be considered as members of the class of things that are stacked up one after the other. The justification for this assumption rests, not on its self-evident truth, but on an elaborate *theory* of sedimentary rocks which in turn rests upon physical and chemical theory. The attempt to explain the movement of blocks of the earth's crust in alpine thrusts belongs to the second phase of historical inference and invokes the explicit application of laws and theories which are, within the context of the inference, regarded as unproblematic.

The tendency to see two distinct levels of inference is quite characteristic of historical discipline. It manifests itself in discussions of historical explanation which consider how historical events are explained and do not consider how historical events are obtained in the first place. It manifests itself also in discussions of the bearing of paleontological evidence upon the validity of evolutionary theories which do not take into account the extent to which the presupposition of those theories might condition that evidence.

I suggest that geologists find Gilbert's almost wholly particularized account of geological inference to be a satisfactory treatment of what might be called "primary historical inference." Geologists apparently do not regard this kind of inference as resting on theoretical, or even general, principles. According to them, however, higher level historical inferences do.

In the mainstream of geology, the scientific character of the discipline has been assured by its demonstrable connections with physical theory. Geologists have not been able to explicate this connection in the concepts of philosophical analysis, but they have been able to secure and utilize the connection scientifically. The geology of the late nineteenth and twentieth centuries represents a triumph of the directional view of earth history. The highly directionalist concept of continental drift and seafloor spreading have not been developed *in spite of* an immutable natural order, they have rather been developed in necessary connection with the most explicit expression of that order, contemporary physical theory. We might describe the geologists of the last one hundred years in Ryle's (1949, pp. 7-8) words: "They are like people who know their way about their own parish, but cannot construct or read a map of it, much less a map of the region or continent in which their parish lies." The lack of such a map has not hampered geologists at all. Their account of a directional earth history within the context of a self-imposed rational order stands as one of the great intellectual achievements of our time.

### References Cited


Basalla, G., Coleman, W., and Kargon, R. K., 1970, Victorian science: Garden City, New York, Doubleday and Company, 510 p.

Gilbert, G. K., 1886, The inculcation of scientific method by example, with an illustration drawn from the Quarternary geology of Utah: Am. Jour. Sci., 3rd ser., v. 31, p. 284-299.

Hubbert, M. K., and Rubey, W. W., 1959, Role of fluid pressure in mechanics of overthrust faulting. I. Mechanics of fluid-filled porous solids and its application to overthrust faulting. Geol. Sci. America Bull., v. 70, p. 115-166.

Leet, L. D., and Judson, S., 1954, Physical geology: New York, Prentice-Hall, 466 p.

Lovejoy, A. O., 1960, The great chain of being: New York, Harper and Row, 376 p.

Reade, J. M., 1908, The mechanics of overthrusts: Geol. Mag., new ser., dec. 5, v. 5, p. 518.

Rudwick, M. J. S., 1970, The strategy of Lyell's *Principles of Geology*: Isis, v. 61, p. 5-33.

Ryle, G., 1949, The concept of mind: New York. Barnes and Noble, 334 p.

Simpson, G. G., 1963, Historical science, *in* Albritton, C. C., ed., The fabric of geology: Reading, Mass., Addison-Wesley, p. 24-48.

Smoluchowski, M. S., 1909, Some remarks on the mechanics of overthrusts: Geol. Mag., new ser., dec. 5, v. 6, p. 204-205.

CHAPTER SIX

# Continental Drift and Scientific Revolution

## INTRODUCTION

It was inevitable that the geologists and geophysicists who contributed to the formulation of the "new global tectonics" should have discovered Thomas Kuhn. In 1962 Kuhn published his *The Structure of Scientific Revolutions*. This is not far from the date that earth scientists have come to mark as the beginning of a particularly significant decade. It now is held generally that during this decade a "scientific revolution" occurred. Some authors, notably Wilson (1968), Cox (1973), and Hallam (1973), have invoked Kuhn's theory of the history of science explicitly in their treatment of recent events in geology. These works are valuable historical accounts of a remarkable period of intellectual ferment. I find Cox's treatment particularly exciting and illuminating. But the doctrines of Kuhn serve these authors as little more than a source of rationalization for the opinion that a scientific revolution has occurred in earth science. I am concerned that in uncritically labeling recent events in earth science as a revolution, and by implication a "Kuhnian revolution," we may miss something significant about the history of geology and, more importantly, something fundamental about the very nature of geologic knowledge. I shall, therefore, examine with some care the claim that geology is just now in the midst of a scientific revolution.

## PARADIGMS AND NORMAL SCIENCE

### *A*

Kuhn rejected the view that science develops by the piecemeal addition of items to the stockpile of scientific knowledge. He suggested rather that periods of "normal science," which are characterized by theoretical stability, are broken occasionally by episodes of rather sudden change in theoretical foundation. Kuhn presented the essentials of this view early in *The Structure of Scientific Revolutions.* He wrote (pp. 10-11),

> In this essay, "normal science" means research firmly based upon one or more past scientific achievements, achievements that some particular scientific community acknowledges for a time as supplying the foundation for its further practice. Today such achievements are recounted, though seldom in their original form, by science textbooks, elementary and advanced. These textbooks expound the body of accepted theory, illustrate many or all of its successful applications, and compare these applications with exemplary observations and experiments. Before such books became popular early in the nineteenth century (and until even more recently in the newly matured sciences), many of the famous classics of science fulfilled a similar function. Aristotle's *Physics*, Ptolemy's *Almagest*, Newton's *Principia* and *Opticks*, Franklin's *Electricity*, Lavoisier's *Chemistry*, and Lyell's *Geology*—these and many other works served for a time implicitly to define the legitimate problems and methods of a research field for succeeding generations of practitioners. They were able to do so because they shared two essential characteristics. Their achievement was sufficiently unprecedented to attract an enduring group of adherents away from competing modes of scientific activity. Simultaneously, it was sufficiently open-ended to leave all sorts of problems for the redefined group of practitioners to resolve.
>
> Achievements that share these two characteristics I shall henceforth refer to as "paradigms," a term that relates closely to "normal science." By choosing it, I mean to suggest that some accepted examples of actual scientific practice—examples which include law, theory, application, and instrumentation together—provide models from which spring particular coherent traditions of scientific research.

Wilson, Cox, and Hallam held that the hypotheses of continental drift and plate tectonics constitute a paradigm which recently has been accepted by the geologic community. This suggestion seems plausible enough; but before accepting it, let us take a careful look at the whole body of geologic knowledge with the paradigm concept in mind.

*B*

There is an immensely complicated body of knowledge shared by geologists, a significant part of which is not expressed explicitly. There is, however, an important part of that body of knowledge that is identified easily. It consists of what geologists, and all others in the scientific community, consider to be the most fundamental and comprehensive scientific principles of their time. Scientific principles are formulated explicitly in theories. A theory, in this technical sense, is a set of statements of general form which, together with statements about initial and boundary conditions, provide a basis for the logical derivations of a great number and variety of singular descriptive statements. General theories thus cover or comprehend particular events. Geologists do not set out to verify or falsify theories. The laws of physics are not questioned within the context of geologic inference. They are simply presupposed. Geologic events must be such as not to falsify fundamental theories. When an inferred event appears to be forbidden by the laws of physics, the resulting "paradox" is not resolved by altering the laws of physics to accommodate the event, but by altering the event to accommodate the laws of physics. A striking example of this inferential stratagem is to be found in the attempt to resolve the "paradox of major overthrusts" (for a discussion of this case see pp. 79-82).

Physical theory serves as a conceptual framework which the geologist does not and, according to the well-understood rules of his game, cannot transcend. In accepting an unproblematic background of theoretical propositions geologists are prevented from assuming that any logically conceivable event whatever may occur. To impose some restrictions upon what we are permitted to say about the world is simply to engage in rational discourse. The acceptance of physical theory provides the necessary protection against chaos in geologic discourse and at the same time frees geology from the stranglehold of "pure" observation. Observations alone cannot tell us what is possible, they can only tell us what is. If geologists were to invoke only observations and the empirical generalizations based upon them in their historical inferences, then the past would turn out to be exactly like the present. Under the umbrella of comprehensive theories we may consider what is possible. This is all that makes the pursuit of history interesting, for it is all that permits us to imagine that the past was in any significant way different from the present.

I have in the above remarks propounded a version of that venerable geologic doctrine, "the uniformitarian principle." It is an explicit statement of a view that is held, I believe, by most geologists. What I have said, therefore, is descriptive rather than prescriptive.

*C*

The concept of the paradigm is designed by Kuhn to enhance our understanding of the growth of science. For scientific growth to occur, according to Kuhn, paradigms must be doubted, attacked, and ultimately repudiated. Geologists never doubt, attack, or repudiate physical theory. Among geologists there is unshakable consensus concerning what is to be regarded as inviolate within the context of geologic inference. For them the fundamental laws of physics do not function as empirical statements subject to observational test, but rather as rules not to be violated under any circumstances (for a discussion of Newton's second law as a rule see Hanson, 1958, p. 100). Consider Kuhn's remarks on normal theoretical work (1962, p. 30):

> A part of normal theoretical work, though only a small part, consists simply in the use of existing theory to predict factual information of intrinsic value. The manufacture of astronomical ephemerides, the computation of lens characteristics, and the production of radio propagation curves are examples of problems of this sort. Scientists, however, generally regard them as hack work to be relegated to engineers or technicians. At no time do very many of them appear in significant scientific journals. But these journals do contain a great many theoretical discussions of problems that, to the non-scientist, must seem almost identical. These are the manipulations of theory undertaken, not because the predictions in which they result are intrinsically valuable, but because they can be confronted directly with experiment. Their purpose is to display a new application of the paradigm or to increase the precision of an application that has already been made.

There is a good deal of work carried on by geologists that may appear to be of the sort described by Kuhn. But the theoretical investigations of the geologists are not undertaken to confront the accepted theory with experiments and thus to add to its empirical support, but rather to use the theory in new roles in historical investigation. A dramatic example is provided by the application of the principles of isotope chemistry to paleoecology. This work was not undertaken to bolster any paradigm. It was undertaken to permit a new and significant kind of historical inference. In historical inferences events are regarded as ends in themselves rather than as a means of getting at theories. However deeply the geologist becomes involved in the theoretical aspects of a current physical paradigm, his goal is not to verify or falsify the paradigm. His work is not directed at the paradigm at all. For him, physical theory is accepted unquestioningly in order that the primary business of historical inference may proceed.

The body of shared knowledge and belief may be in large part identical for geologists and physicists. Most physicists, like geologists, are not involved in any attempt to alter the fundamental theory of their day. But if Kuhn's paradigm is to figure in explanations of scientific growth, there must be from time to time a handful of "physicists" who are willing to attack it. What may be a paradigm for physical science is a kind of inviolable "superparadigm" for geology.

Again, I must make sure the reader understands that I am attempting to describe the character of geology. I make no claim that geology must be as I have described it, or that it should be. It is conceivable in principle that a geologist confronted with a paradox resulting from the incompatability of a geologic event and physical theory might respond by insisting that physical theory be altered to bring it into accord with the geologic event. Paleontologists sometimes fault biologic theory because it is incompatible with the record of biologic events. Geologists never fault physical theory because it is incompatible with the record of physical events. The reason for this intriguing difference between paleontologists and geologists is probably as much a historical and sociological question as it is a methodological one and is, in any case, beyond the scope of this discussion to consider.

Geologists have had no role in the revolutions which have led to the overthrow of comprehensive theoretical paradigms. When a theoretical paradigm is abandoned in favor of another it may have revolutionary consequences for geology. The revolution that led to the overthrow of classical mechanics in the late nineteenth and early twentieth centuries affected geology in an important way, but it was not accomplished with the participation of geologists or at their instigation.

## CONTINENTAL DRIFT

### *A*

All this is preliminary to a discussion of a current revolution in geology. Geologists are not talking about a change in theoretical paradigm taking place outside geology. They are concerned with a change going on inside geology which leaves the theoretical paradigm unscathed.

Hallam (1973, p. 106) wrote, "A paradigm is an accepted model or pattern of beliefs but it is easier to explain by example than define. Kuhn draws his examples from physics and chemistry." It is no mere accident of Kuhn's background and academic training that he chooses his examples of paradigms from chemistry and physics. It is clear that for Kuhn, paradigms exercise their pervasive influence by virtue of their being general knowledge systems. He recognizes different degrees of comprehension, but he does not consider any hypothesis which is concerned wholly with particular events. The hypothesis of continental drift *is* concerned wholly with particular events. It does not consist of a body of general propositions of which events must be instances, but is itself an instance which must be consistent with some body of general propositions. It is not formulated in universal or even general terms, but in singular descriptive terms. It makes no assertions about an untimebound and unspacebound natural order, but about conditions prevailing at particular times and places. It is, in short, historical rather than theoretical. This important distinction was recognized by Chamberlin who wrote (1928), "Wegener's hypothesis is no general theory of earth behavior or earth deformation. It describes simply one supposed breaking up of a consolidated land mass and the migration of the different fragments."

In its most highly developed state the hypothesis of continental drift would consist of an exhaustive description of a nonrepeatable succession of historical events. "Continent" hardly functions as a general term in the hypothesis. To put something in the class of continents permits us to predicate very little of that thing—little more, in fact, than that it is capable of motion with respect to others of its kind. What is significant is not what we can say of a continent by virtue of its being a member of the class of continents, but what we can say about its motion within some designated coordinate system.

A general law containing the notion of continental drift could, in principle, be formulated. A geologist might be willing to say, for example, "Wherever there are continents, be they on earth or some other planet, they will move with respect to one another." This is not what is meant, however, by "the theory of continental drift," and it is not in this form that the concept figures in geologic inferences.

The singular character of the hypothesis is clear in Wegener (1924, pp. 1, 2), who wrote,

> He who examines the opposite coasts of the South Atlantic Ocean must be somewhat struck by the similarity of the shapes of the coast-lines of Brazil and Africa. Not only does the great right-angled bend formed by the Brazilian coast at Cape San Roque find its exact counterpart in the re-entrant angle of the African coast line near the Cameroons, but also, south of these two corresponding points, every projection of the Brazilian side corresponds to a similarly shaped bay in the African, and conversely each indention in the Brazilian coast has a complementary protuberance on the African. Experiment with a compass on a globe shows that their dimensions agree accurately.
>
> This phenomenon was the starting point of a new conception of the nature of the earth's crust and of the movements occurring there; this new idea is called the theory of displacement of continents, or, more shortly, the displacement theory, as its most prominent component is the assumption of great horizontal drifting movements which the continental block underwent in the course of geologic time and which presumably continue even today.

Wegener has explained the present configuration of the South American and African Atlantic coastlines by supposing that the continents had been joined at some time in the past and had since drifted apart. The hypothesis of continental drift does not serve the function of covering generalization in this explanation, but of initial and boundary conditions. Then what are the unstated generalizations involved? One of them must be a statement to the effect that if two complex irregular physical boundaries may, in principle, be fitted together, then it is more plausible to assume that they were initially coincident than to assume that they just happen to have the same configuration. One might employ this generalization in support of the conclusions that two pieces of paper had been torn from a single sheet.

A statement describing a long-range historical trend is sometimes called a "law." Popper (1957) and others have pointed out that these so-called "laws of historical succession" or "laws of evolution" are not natural laws in the usually accepted sense of this term, because they consist of descriptions of sequences of events. The processes, Popper maintains, may proceed in accordance with certain laws, but the description itself is not a law but a finite conjunction of singular statements.

To claim that the hypothesis of continental drift is historical rather than theoretical is not to claim that it is trivial or to claim that it is insignificant in the recent history of geology. But whatever its significance may be, it cannot be identical to the significance of a theory. And however sudden the shift in opinion concerning it, that shift cannot be just like the changes in opinion concerning theories that figure so importantly in Kuhn's treatment of the history of science. Having emphasized and, I hope, established this point, I should like now to consider the way in which the hypothesis of continental drift has in fact exerted its profound influence on the earth sciences.

## A HISTORICAL PARADIGM

The hypothesis of continental drift will be made to conform to the unproblematic background of physical theory or it will be abandoned. The objection has been raised repeatedly that to suppose the continents moved is a violation of physics. Some geologists rejected continental drift on these grounds, whereas others proceeded to alter initial and boundary conditions in such a way as to permit it to occur. No one cited the lack of agreement between the historical event and physical theory as grounds for rejecting or altering the theory. To quote again from the well-known AAPG volume of 1928, Longwell wrote (p. 152):

> It is obvious that the results of geophysical examination, so far as they go, are generally unfavorable to the displacement hypothesis, but they are not conclusive. In fact the geophysicists turn the problem back to us, with the statement that geologists alone can determine whether the geophysical forces have had "in geologic history an appreciable influence on the position and configuration of our continents." Let us therefore re-examine some of our evidence, to see whether it is compelling. If it is, then physical geologists should be content to accept the fact of displacement, and leave the explanation to the future.

A fully satisfactory explanation of continental drift has not yet been formulated. The attempt to achieve such an explanation goes on because by now western European and American geologists are, by and large, convinced that the evidence for continental drift is indeed compelling. The task that geologists and geophysicists have set for themselves has many aspects of what Kuhn has called "puzzle solving." He wrote (1962, p. 36), "Bringing a normal research project to a conclusion is achieving the anticipated in a new way, and it requires the solution of all sorts of complex instrumental, conceptual, and mathematical puzzles. The man who succeeds proves himself an expert puzzle-solver, and the challenge of the puzzle is an important part of what usually drives him on."

Kuhn made clear that for him puzzle solving is not a routine undertaking of little significance in the history of science. It is an important activity requiring imagination and ingenuity on the part of the investigator. But I am reluctant to call the attempt to accommodate continental drift and physical theory puzzle solving, because it might not do justice to the profound significance of the hypothesis in geology. That significance lies in the fact that the hypothesis of continental drift, despite its singular character, has a unifying effect. Theories unify by covering a great number and variety of phenomena. By invoking the theory of mechanics, planetary motion, falling bodies, and major overthrusts may all be explained. The hypothesis of continental drift, being particular in form, cannot cover phenomena. It must itself be covered. Its unifying function lies in the fact that it figures in many explanations as a particular antecedent condition. The distribution of animals, the remanent magnetism of rocks, and the shapes of the continents can all be explained, it is claimed, by supposing as an initial condition that certain continental masses formerly were joined and have broken apart and drifted. It is not simply a matter of explaining phenomena with which we are already familiar. Once a hypothesis has been proposed, it directs the attention of scientists to certain kinds of phenomena, this being true whether the scientists are trying to verify or falsify the hypothesis. There is no question, for example, that paleomagnetic studies have received their main impetus from the hypothesis of continental drift.

Because the hypothesis of continental drift is itself something to be explained, it unifies at another level. Munyan wrote (1963, p. 2),

> The implications of this proposal [continental drift] seriously challenged many of the beliefs and theories of the constitution of the earth, its physical properties, tectonics, and biological developments. As a result a considerable furor of opposition arose on all counts, but, in particular, the geophysicists alleged that drift was out of the question because the crust could not endure such forces.

The physical properties of the earth are discovered by attempting to construct them in such a way as to permit all the phenomena which we observe and infer. Continental drift is one phenomenon which has been inferred and consequently presents an opportunity for geologists to extend their knowledge of the earth's physical properties. For one convinced that continental drift has occurred, the earth must have physical properties which permit it.

The hypothesis of continental drift unifies in two ways. It figures as a common antecedent condition in a great variety of explanations, and attempts to explain it significantly influenced the direction of investigation. The effect of the hypothesis of continental drift upon geology has not been exaggerated. A substantial proportion of geologists in North America is engaged in an attempt to verify, falsify, or explain continental drift. The hypothesis of continental drift constitutes a kind of paradigm. It is not like the theoretical paradigms of Kuhn. It is the sort of paradigm particularly appropriate to the most scrupulously historical of all sciences, in that it is itself historical. It imposes upon those who accept it a particular version of history rather than a general theory of the world.

## PLATE TECTONICS

In the attempt to bridge the inferential gap between physical theory and the historical hypothesis of continental drift, certain hypotheses have been recognized as particularly significant and have themselves been accorded the status of "theories." Such a hypothesis is "plate tectonics."

Cox (1973) boldly formulated the hypothesis of plate tectonics in a series of definitions and postulates. His Postulate I states (p. 41), "The plates are internally rigid but are uncoupled from each other. At their boundaries two plates may pull apart or slip one beneath the other, but within the plates there is no deformation." But, Cox continued (p. 43),

> Of course, every geologist knows that these postulates do not provide a complete explanation for all geologic information—if the continents were rigid blocks, there would be no need to study structural geology. Yet the conclusions drawn from plate tectonics offer rational explanations for so many of the earth's major tectonic features that the basic assumption appears to be justified with only two minor modifications: *most* large-scale deformation occurs in narrow zones between plates that are *nearly* rigid.

It may appear at first glance that the hypothesis of plate tectonics is no more general than the hypothesis of continental drift. It describes the particular behavior of individual objects. But plate tectonics cannot, like continental drift, be explicated fully in a detailed historical account.

There is in plate tectonics a crucial theoretical, and therefore general, dimension which is not reducible to a description of events. That theoretical dimension is not provided by a geologic hypothesis formulated within the last decade. It comes from the familiar and inviolable "superparadigm." By virtue of something being a member of the class "plates," we can attribute certain theoretical properties to that thing; properties such as rigidity and mass. Providing plates with properties of significance within physical theory permits geologists to invoke that theory with all its immense inferential efficacy in their discussions of plates and continental drift.

## REVOLUTION

Up to this point I have discussed continental drift as a ruling hypothesis in geology. I have said nothing about how it came to be accepted so widely. It is too soon to attempt to write a history of geology and geophysics during the past fifty years. But in pursuing my purpose I should like to consider briefly the radical shift in opinion about continental drift in the light of Kuhn's theory of scientific growth. According to Kuhn (1962, pp. 52-53),

> Discovery commences with the awareness of anomaly, *i.e.*, with the recognition that nature has somehow violated the paradigm-induced expectations that govern normal science. It then continues with a more or less extended exploration of the area of anomaly. And it closes only when the paradigm theory has been adjusted so that the anomalous has become expected.

The recognition of anomaly does not, by itself, entail revolution. Anomalies may be regarded as puzzles rather than as counterinstances for a paradigm. Gravitational theory is not abandoned because a heavy object rises upon being released. An attempt is made instead to show that a trick has been performed which has no force as a falsifying event. The conventional account of what Wilson (1968) has suggested we call the "Wegenerian revolution" holds that the evidence for continental drift has become so overwhelming that no reasonable man any longer can ignore it. Geomagnetism has been singled out as providing an especially compelling body of evidence. It is a source of counterinstances to the hypothesis of the fixity of continents. But as Kuhn pointed out in the following passage (1962, p. 79), counterinstances to every theory may be recognized.

> Excepting those that are exclusively instrumental, every problem that normal science sees as a puzzle can be seen, from another viewpoint, as a

counterinstance and thus as a source of crisis. Copernicus saw as counterinstances what most of Ptolemy's other successors had seen as puzzles in the match between observation and theory. Lavoisier saw as a counterinstance what Priestley had seen as a successfully solved puzzle in the articulation of the phlogiston theory. And Einstein saw as counterinstances what Lorentz, Fitzgerald, and others had seen as puzzles in the articulation of Newton's and Maxwell's theories. Furthermore, even the existence of crisis does not by itself transform a puzzle into a counterinstance. There is no such sharp dividing line. Instead, by proliferating versions of the paradigm, crisis loosens the rules of normal puzzle-solving in ways that ultimately permit a new paradigm to emerge. There are, I think, only two alternatives: either no scientific theory ever confronts a counterinstance, or all such theories confront counterinstances at all times.

Events do not identify themselves intrinsically as being either puzzles or counterinstances. The distinction can be made only from some theoretical vantage point. During the first half of the 20th century the driftists confronted geology with states and events which they considered to be counterinstances to the hypothesis of the fixity of continents. The opponents of drift regarded these events as puzzles to be accounted for within the context of the fixity of continents. The configuration of the coast of the South Atlantic, which figures so importantly in the arguments of the driftists, was not a counterinstance that entailed the abandonment of the fixity of continents. It wasn't even much of a puzzle. It was an accident.

By the middle of the 1950s American and western European geologists began to accept continental drift in increasing numbers. What had been considered puzzles for the hypothesis of the fixity of continents became counterinstances. Perhaps it was at this time that the crisis, which Kuhn considered to be a necessary prerequisite to revolution, occurred. I do not wish to dwell upon this for the reason given previously. We geologists cannot at this time stand off and dispassionately examine a movement in which we are so intimately and enthusiastically involved. By the middle of the 1960s, in any case, the western geologic community had accepted a new conceptual scheme to "define the legitimate problems and methods of a research field for succeeding generations of practitioners" (Kuhn, 1962). For Anderson (1972), to cite just one example, the crisis is over. Anomalously low heat flow on the flanks of slow-spreading midocean ridges is no counterinstance to the sea-floor-spreading hypothesis. It has nothing whatever to do with testing the hypothesis. It is a puzzle to be solved within the context of that hypothesis. In solving the puzzle, or in resolving the paradox, Anderson discovered something of geologic interest.

## CONCLUSION

History is written within a conceptual framework which imposes a limit upon what we may suppose to have occurred in the past. To accept such a limit is simply to operate within the realm of rational discourse. In any historical discipline questions arise as to how much restriction should be imposed upon past events. The great debates of the nineteenth century concerning the uniformitarian principle were directed to this question. By the middle of the twentieth century the question had been tacitly answered. Geologists would be guided in their historical deliberations by the most comprehensive principles expressed in the chemical and physical theory of their day. The acceptance of this superparadigm, while guarding against chaos in historical discourse, provided a means of escaping the limitation to "what can actually be observed" that so many geologists sought to impose. Theory permits the geologists to decide what is possible and what is not. But history goes beyond a consideration of what is possible to a consideration of "what actually happened." Physical theory exerts its pervasive influence because geologists have agreed that no matter what the events of history appear to be, an effort must be made to bring them into accord with what physical theory dictates as possible. This task is accomplished by altering the specific conditions and events of the past to secure such accord. An account of the conditions and events of the past is a historical chronicle. The attempt to resolve paradoxes is not some dramatic but incidental activity. It occupies a central role in geologic methodology.

Changes in physical theory accomplished outside of geology may effect the character of historical inference in a dramatic way. But geologists have a way of altering the history of the earth without tampering with the theoretical apparatus which they presuppose. When the boundary conditions introduced during geologic inference are temporally and spatially widespread, and when they are comprehensive in the sense that they are invoked in many explanations, they take on the characteristics of a paradigm. It is just the sort of paradigm that we would expect to be developed within a discipline so radically historical as geology.

It has not been my intention to provide yet another consideration of the adequacy of Kuhn's views as a general theory of the history of science, but only to consider their applicability to the recent events in geology. In my judgment Kuhn's theory illuminates the last decade in the history of geology. It should be pointed out, however, that the acceptance of continental drift and plate tectonics, although having many of the aspects of a revolutionary change, might as well be regarded as the latest step in an evolutionary movement away from a cyclic view of earth history which began at least as long ago as the middle of the nineteenth century. This latest step permits geologists to give an account of earth history as unidirectional, irreversible change within the confines of an immutable natural order which finds its most explicit expression in physical theory.

## References Cited

Anderson, R. N., 1972, Petrologic significance of low heat flow on the flanks of slow-spreading midocean ridges: Geol. Soc. America Bull., v. 83, no. 10, p. 2947-2956.

Chamberlin, R. T., 1928, Some of the objections to Wegener's theory, *in* Theory of continental drift: Am. Assoc. Petroleum Geologists, p. 83-87.

Cox, A., 1973, Plate tectonics and geomagnetic reversals: San Francisco, W. H. Freeman, 702 p.

Hallam, A., 1973, A revolution in the earth sciences: Oxford, Clarendon Press, 127 p.

Hanson, N. R., 1958, Patterns of discovery: Cambridge, Cambridge Univ. Press, 240 p.

Kuhn, T. S., 1962, The structure of scientific revolutions: Chicago, Univ. Chicago Press, 172 p.

Longwell, C. R., 1928, Some physical tests of the displacement hypothesis, *in* Theory of continental drift: Am. Assoc. Petroleum Geologists, p. 145-157.

Munyan, A. C., 1963, Introduction to polar wanderings and continental drift, *in* Polar wandering and continental drift: Soc. Econ. Paleontologists and Mineralogists Spec. Pub. 10, p. 1-3.

Popper, K. R., 1957, The poverty of historicism: Boston, Beacon Press, 166 p.

Wegener, A., 1924, The origin of continents and oceans: London, Methuen, 212 p. (Translated by J. G. A. Skerl from 3d German edition; published in 1922.)

Wilson, J. T., 1968, Static or mobile earth—the current scientific revolution: Am. Philos. Soc. Proc., v. 112, no. 5, p. 309-320.

CHAPTER SEVEN

# Geologic Time

THE ORDER OF EVENTS

We know what time is. We live in it. But historical time is not given in immediate experience and cannot be regarded as an extension of common sense time into the past. It has its own special properties which are imposed upon it by the assumptions and procedures which we use to construct it.

In one phase of historical investigation geologists describe the present state of the earth. The term "present" here does not refer to an instant in time but to an interval which may cover tens or even hundreds of years. The extension of the geological *specious present* is justified by the fact that objects of geological significance remain relatively unchanged as compared to other objects. But even during the geological present, significant change occurs and description of change contains essential reference to time. The "before" and "after" and "at the same time" of these descriptions often refer to temporal relationships among states in the awareness of an observer. Observation statements describing processes, even though they denote temporal relationships, should not be confused with descriptions of historical events and processes. Historical events, and the temporal relationships among them, must be inferred. The inference rests upon the conviction that the events of the present occur because of a causal relationship to the events of the past. The connection between an event in the present and an event in the past is justified by adducing geological generalizations which point to a necessary relationship between them. Geological explanations consist of temporally ordered sequences of events. The theory of geology may consequently be regarded as, among other things, a set of temporal ordering principles. Any two events, no matter what the distance between them, may be ordered in time provided a causal connection between them may be inferred. At this point I wish only to consider events relatively close together spatially or, in familiar geological terms, events which can be inferred from a single local section. The formation of a body of rock is explained by pointing to some antecedent events which, it is held, were causally related to one another. By virtue of their causal relationship they are temporally ordered.

Events may be ordered temporally by means of the law of superposition without assuming any causal connection between them. Superposition, unlike causal connection, can only be used to order events which are very close to one another in space. A pair of events ordered causally or in superposition is ordered in the relationship "earlier than —later than." The local ordering of biological events rests ultimately upon the ordering of physical events and upon the principle, so clearly stated and amply justified by Steno, that certain objects contained in rocks are the remains of organisms which existed just before the formation of the sediments in which they were enclosed.

### THE MEASUREMENT OF TIME

The ordering of events in the relation "earlier than—later than" is a topological operation. The principles upon which this ordering is based contain no reference to a metric and consequently cannot serve to justify the assignment of numerical values to the interval between events. In short, they cannot provide a basis for the *measurement* of time.

Intervals of time may be measured by comparing them with the uniform intervals marked off by a clock. How do we know that the intervals defined by any clock are uniform? In answer to this question, Reichenbach stated (1958, p. 116), "There is only one answer: we cannot test it at all. There is basically no means to compare two successive periods of a clock, just as there is no means to compare two measuring rods when one lies behind the other. We cannot carry back the later time and place it next to the earlier one." And further (same page), "A solution is obtained only when we apply our previous results about spatial congruence and introduce the concept of a *coordinative definition* into the measure of time. The equality of successive time intervals is not a matter of *knowledge* but a *matter of definition*." A coordinative definition consists, in effect, of adopting some periodic process as a standard clock. Does not this remarkable fact about the nature of clocks open the way for the adoption of clocks especially suited to the measurement of geologic time? It has, after all, been suggested that certain geological events are periodic. Suppose that someone, upon being asked to justify his contention that the intervals of time between a series of geological events—for example between orogenies—are uniform, were to state simply, "It is true by definition."

This might give us pause to reflect upon the nature of clocks, but we would reject the occurrence of orogenies as periodic nonetheless. A physician determines that a pulse beat is irregular by comparing it with the intervals defined by his wristwatch. Even though the intervals defined by the second hand are uniform by definition only, he would not hesitate to reject a patient's claim that his pulse was periodic. Why is it, then, in view of the definitional character of the uniformity of time intervals, that certain successive events are accepted as standards for measurement and others are rejected? The answer is that clocks which are consistent with the "fundamental" scientific theories are accepted as standards. A clock, though it permitted simple geological or physiological descriptions, would be rejected if its adoption necessitated a change in the laws of physics. This is not to say that we are restricted to a single kind of system in the construction of clocks. Clocks based upon a number of different physical systems yield more or less consistent estimates of duration. If, for example, a spring clock and a pendulum clock are used to measure the interval of time between two events, they will yield consistent values if certain well-understood precautions have been taken. To determine whether two systems do in fact allow consistent estimates of duration is a matter for empirical or theoretical test.

There is little evidence, either theoretical or empirical, that sequences of geological events define uniform intervals of time as compared to standard physical clocks. In rare cases a geologist may have good reason to suppose that measurements based upon a succession of geological events will be consistent with measurements based upon a standard clock. Varves and tree rings are, for example, regarded as reliable standards for the measurement of time, because the processes thought to lead to their formation can be observed and their periodicity measured against standard clocks and, even more significant, because these processes are held to bear a rather simple and direct relationship to periodic events within the solar system.

A hypothesis of the periodicity of events need not rest upon direct measurement. Justification for the contention that fluctuations in the obliquity of the ecliptic are periodic is not to be sought for in direct measurement, but can be theoretically justified within mechanics. If it could be shown that changes in the obliquity of the ecliptic resulted in geologically detectable events, then the basis for a geological clock would be provided, even though the periodicity of these events could not be measured directly.

Clocks based upon *closed* periodic systems provide the most satisfactory standards for the measurement of time. Closed systems are those whose behavior is free from the effect of forces external to the system. Reichenbach (1958, p. 119) points out that because no system is completely closed, we can regard systems as being closed only to a *certain degree of approximation.* In regard to its diurnal rotation the earth forms a nearly closed system, and for this reason a clock based upon its period of rotation is considered to be highly reliable. Physical and geological theory require that we regard geological systems as badly closed. This is, in itself, a sufficient basis for the rejection of almost any "clock" based on a geological system.

It is possible to measure intervals of time by nonperiodic processes. For such a process to serve as a satisfactory clock it must be closed to a high degree of approximation and its use must lead to measurements which are consistent with those obtained from standard clocks. The decay of radioactive elements, according to physical theory, fits these criteria. The suggestion is sometimes made that the process of organic evolution, like that of radioactive decay, can be used to measure intervals of time. Teichert, for example, says (1958, p. 116),

> Measurements of past time periods depend on observational methods basically similar to those applied in the measurement of contemporaneous time. They are deductions from observations of rock properties or structures (including fossils) which may be interpreted as having resulted from the operation of processes which are either unidirectional and irreversible or occur with a fixed periodicity. The only terrestrial phenomena which fulfill these conditions and have proceeded universally for considerable periods are radioactive disintegration and the evolution of life.

To draw an analogy between the role of radioactive disintegration and organic evolution in geological inference is apparently tempting, for Teichert is by no means the only geologist to have done it (see, for example, Donovan, 1966, among many others). In the paper from which the above quotation was taken Teichert says (p. 102),

> I do not wish to suggest that in the biological world genes mutate with the predictable precision of radioactive decay. Evolution is continuous only in the sense that life itself is a continuously operating chain of events. There are discontinuities and jumps, and it is these that are often particularly useful and valuable in the paleontological record. However, the overall effect of evolutionary processes which determine the composition and character of life communities was felt all over the world at the same time and at the same rate. On this uniformity of the over-all result depends paleontological correlation, a highly empirical science that can never take its own results for granted and yet has succeeded in accumulating a huge body of almost infallible guides to the interpretation of the geologic past.

In the statement, "I do not wish to suggest that in the biological world genes mutate with the predictable precision of radioactive decay," Teichert clearly rejects the claim that any identifiable sequence of biological events could be said to define uniform intervals of time. There is indeed no empirical or theoretical reason to suppose that this is the case. Organic evolution, therefore, cannot serve as a basis for assigning numerical values to intervals of time. Then what does Teichert mean when he suggests that we can measure time by the evolution of life? In an attempt to answer this question let me suggest a model with the essential features of Teichert's evolutionary "clock." Imagine a machine which strikes a single note at intervals which are known to be nonuniform with respect to standard clocks. Each note in the sequence differs in tone from every other note, and each time the machine strikes it records a mark on a tape which is distinguishable from every other mark made by the machine. There are a number of these machines at different points in space and they are synchronized, that is to say when one strikes they all strike at the same time, with a note of the same tone, and leave a structurally identical mark on a tape. Such a set of machines, which I submit represents a mechanical analogue with all of the essential features of Teichert's "evolution of life," would be extremely useful for extending temporal ordering from one place to another, but it could not serve as a basis for the measurement of time.

## ABSOLUTE AND RELATIVE TIME

When a geologist talks about "relative time" he is referring to the order of events. The sense in which order is relative is perfectly clear. If a set of events is temporally ordered, then the elements of the set stand to one another in the *relation* "earlier than—later than," or the *relation* "simultaneous with." The term "absolute time" in geology refers to the length of time between events. The sense in which the measurement of intervals is absolute is not at all clear. Measurement is relative at least in the sense that it requires reference to some standard. There is an obvious and fundamental distinction to be made between the ordering of events in time and the measurement of the time intervals between them, which does not in any way hinge upon our understanding of the sense in which measurement is supposed to be absolute. The measurement of time intervals consists of assigning numerical values to them. The ordering of events need involve no quantification whatever. The significant distinction is between metric and nonmetric.

But there is probably more than just the notion of measurement that lurks behind "absolute time." Eighteenth-century geology, after all, borrowed the term from classical physics. Perhaps contemporary geologists mean to use it in its full Newtonian sense. Newton defines absolute time in the *Principia* in the following way:

> 1. Absolute, true, and mathematical time, of itself, and from its own nature, flows equably without relation to anything external, and by another name is called duration: relative, apparent, and common time, is some sensible and external (whether accurate or unequable) measure of duration by means of motion, which is commonly used instead of true time; such as an hour, a day, a month, a year.

In regard to Newton's concept of absolute time Withrow (1961, p. 34) remarks:

> In practice we can only observe events and use processes based on them for the measurement of time. The Newtonian theory of time assumes, however, that there exists a unique series of moments and that events are distinct from them but can occupy some of them. Thus temporal relations between events are complex relations formed by the relation of events to the moments of time which they occupy and the before-and-after relation subsisting between distinct moments of time.

There is no method for discovering the "unique moments" of geologic time, nor is there any way of showing that geologic time "flows." According to Grünbaum (1964, pp. 229-30) "*Physically*, certain states are later than others by certain amounts of time. But there is no "flow" of *physical* time because physically there is no egocentric (psychological) transient *now*." Flow is not a feature of physical time, but of common sense, or psychological, time. The notion of the flow of time is suggested by, as Grünbaum has put it, the steady shift of the "now contents" of our consciousness. How then are we to impose *flow* on geologic time? The answer is that there is no way to impose it on time beyond the reach of immediate experience. *Geologically* "certain states are later than others by certain amounts of time," but there is no flow of geologic time.

Does the idea of Newtonian absolute time enter into the geologist's conception of time? I think it does. It is certainly a significant factor in Jeletsky's conception of geologic time, even though he calls it by another name. He says (1956, pp. 701-2),

> There is a certain advantage in using geologic time units as ideal time units based solely on the unknowable *abstract* time, and independent from all material criteria of geochronological division. So interpreted, the ideal geologic time units would be strictly theoretical; they would serve solely the purpose of keeping us aware of the existence of the uninterrupted *abstract* geologic time behind its partial biological and physical standards.

Jeletsky decides, for various practical reasons, not to define geologic time units in terms of abstract time. It is absurd even to suggest, however, that there might be an advantage in using terms in such a way as to remind us of the *existence* of something which is *unknowable.*

Allusions to Newtonian absolute time are not usually so explicit as in the passage from Jeletsky. I suspect, however, that the notion of a *flowing, unknowable* geologic time lies hidden in many discussions and rises occasionally to make its contribution to the prevailing confusion. In any case, I propose that we abandon once and for all the term "absolute time." There are perfectly familiar terms that can do all the work of "absolute time" and do it better by avoiding the ambiguity attached to this term.

## CORRELATION BY SIGNAL

Inferences of geologic time based upon radioactive decay depend, in principle, upon the measurement of time intervals between events. Not only can the interval of time between events be determined radiometrically but events, either in the same place or in different places, can be ordered by this means. Given three points on a line, the distance from each point to the other points, and the direction of increasing magnitude, the order of the points is uniquely determined. This relationship of distance and order holds equally well for "points" in time and the "distance" between them. Inferring order from time is simplified in geology because one point in time whose distance from other points may be known is the present, and the "other side" of this point, the future, is ignored. All a geologist needs to know to order three points in time if one of the points is the present is the distance from the present to the other points, which is a roundabout way of saying that we can determine the order of two events if we know how long ago they occurred. This method of ordering does not depend either upon a causal connection between the ordered events or upon the principle of superposition. Radiometric determinations permit, in principle, that events be ordered in the relation "simultaneous with" as well as in "earlier than—later than." If two events occurred at a time equidistant from some other event and on the same "time side" of that event, then they occurred at the same time. In principle this method provides the most satisfactory means of establishing the simultaneity of events. Unfortunately a probable error attends radiometric determinations. We can, however, upon occasion talk meaningfully about the probability that two events occurred simultaneously.

Radiometry provides a useful method of extending temporal ordering, but the applicability of this method is, in practice, restricted. Two events may, however, be temporally ordered with respect to one another in the complete absence of knowledge of the length of the time interval between them, or between either of them and any other event, if they are causally connected. A causal connection between two events separated by any appreciable distance is usually a complex one involving what it is convenient to regard as several events. For this reason it is customary to speak of connection by means of causal chains. In geology, no less than in cosmology, the ordering of events separated from one another in space usually depends upon connecting them by a causal chain or *signal* of finite velocity. The transportation of a volcanic ash provides an example of a geological signal. Generalizations may be adduced to support the conclusion that the deposition of an ash is one end of a causal chain, or the terminus of a signal, which was initiated by an explosive volcanic eruption at another place. The signal is not observed. It is inferred just as an astronomer might infer the arrival of a light signal from the condition of a photographic plate. Biological events, like physical events, are causally related to one another. A fosssil assemblage constitutes evidence of a complex biological event which, it must be supposed, was causally related to other biological events. A series of causally related biological events, then, provides a basis for a signal.

We have at our disposal a vast number of well-confirmed geological and physical generalizations which allow us to associate causally *certain kinds of events* with *certain other kinds of events*. It is quite another matter, however, to relate causally *an instance* of one kind of event with *an instance* of another kind of event. A familiar causal generalization supports the conclusion that an ash fall is causally related to some explosive volcanic eruption. But how can one ash fall be associated with one volcanic eruption chosen from all of the eruptions which we know about? The answer to this seemingly trivial, but actually important, question is that particular associations are made on the basis of the well-supported principle that no two pairs of events covered by a causal generalization are precisely alike. A certain ash fall may be associated with one instance of "volcanic eruption" on the basis of the unique mineralogical composition shared by ash shards and a flow basalt. This in no way negates what has been said about the uniqueness of historical events. Each event, or pair of events if we choose to regard them as such, has characteristics in common with other events, but no two events are precisely alike. In a general way history repeats itself. In the present context it is necessary to emphasize that it never does so precisely. Geological events, and consequently the causal chains composed of them, are nonrepeatable. Without this assumption, the recognition of individual signals would be impossible. There is no theoretical reason to suppose that an event should not be precisely repeated. We justify our assumption of nonrepeatability by pointing out that in any geologically significant situation the number of possible combinations of initial and boundary conditions is so high that the probability of precise repetition is vanishingly small.

The nonrepeatability of sequences of organic events must be assumed but is not easily demonstrated. Suppose someone were to claim that two identical paleontological sequences originated during nonoverlapping intervals of time, thus arguing for the repeatability of organic events. The only effective counter argument would be to point out that the two sequences were correlative. But paleontological correlation, which presupposes the nonrepeatability of organic events, could not be invoked unless independent justification for the presupposition could be presented. Contemporary evolutionary theory strongly supports the contention that organic events are contingently nonrepeatable. Organic evolution, while it provides no basis for the measurement of time, does provide a firm foundation for the fundamental assumption upon which paleontological correlation rests; and that assumption is that a biological signal of particular character is transmitted only once in history.

It may be possible under certain circumstances to distinguish similar causal chains from one another on the basis of their unique arrival and departure times. If, for example, a geologist were to encounter two bentonites of identical composition he would regard them as the termini of two separate causal chains if undisputed stratigraphic evidence indicated that one had been formed from ash deposited during Cambrian time and the other from ash deposited during Cretaceous time. It must be recognized, however, that "undisputed stratigraphic evidence" would rest ultimately upon signals distinguished upon grounds other than their unique arrival and departure times.

The contingent nonrepeatability of geological events provides a basis for the recognition of geological-historical signals. In practice it is usually very difficult to delineate a useful signal. Signals, for example, may be confused because they are so similar. Two causal chains, each of which began with local uplift and ended with the deposition of a suite of sediments, might not be distinguishable if the sedimentary suites marking their ends were very similar. No geological causal chain may, moreover, be regarded as a well-closed system. Physical and biological causal chains are subject to external factors which may alter their character during transit. It might be relatively easy, for example, to associate the erosion of an igneous body with the deposition of a certain gravel, but very difficult to associate the erosion of an igneous body with the deposition of a certain clay.

If the direction of a signal is known, then the ordering of two events connected by it in the relation "earlier than—later than" can be immediately inferred. Once a signal has been defined in terms of two or more events, the determination of its direction will rest upon a knowledge of the temporal direction of the processes comprising the signal and upon the assumption that they are irreversible. It is interesting to note that the laws of dynamics, with which geological principles are supposed to be consistent, relate to processes which are reversible. It will be useful to introduce a distinction, made by Grünbaum, between two senses of "irreversible." He says (1963, p. 210):

> There is both a weak sense and a strong sense in which a process might be claimed to be "irreversible." The weak sense is that the temporal inverse of the process in fact never (or hardly ever) occurs with increasing time for the following reason: certain particular de facto conditions ("initial" or "boundary" conditions) obtaining in the universe independently of any law (or laws) combine with a relevant law (or laws) to render the temporal inverse de facto nonexistent, although no law or combination of laws itself disallows the inverse process. The strong sense of "irreversible" is that the temporal inverse is impossible in virtue of being ruled out by a law alone or by a combination of laws.

Geological processes may be regarded as irreversible in the weak rather than in the strong sense. It is restrictive initial and boundary conditions that render geological processes irreversible. There are no laws, for example, which forbid the reversibility of sedimentary processes. Reversal does not occur because the conditions under which it could are never realized.

If two events can be shown to be directly connected by a physical signal, the determination of the direction of transit poses no problems because it simply depends upon a knowledge of some causal generalizations. The determination of the direction of a biological signal is more difficult. If a geologist concludes that two very similar fossil assemblages represent the two ends of a single causal chain, he may invoke his knowledge of phylogeny in an attempt to show the direction of travel. He might say, for example, that assemblage containing the remains of "more advanced" organisms was younger. But our knowledge of phylogeny, unlike our knowledge of causal laws invoked in the determination of physical signal direction, is not independent of our knowledge of historical succession. Underlying it is a nexus of inferences founded upon causal generalizations, which themselves may invoke phylogeny. This is not to suggest that inferences about the direction of transit of biological signals is necessarily circular. It is clear, however, that care must be taken to avoid presupposing the conclusion in such inferences.

By invoking Steno's principle to the effect that organic remains immediately predate the sediments in which they are contained, it is possible to order physical events on the basis of biological signals and, alternately, to order biological events on the basis of physical signals.

### SIGNAL VELOCITY AND SIMULTANEITY

Suppose that a paleontologist is confronted with two sedimentary beds widely separated in space and containing fossil assemblages of nearly identical character. How does he account for the fact that the assemblages are so similar and in the process conclude that the beds which contain them were deposited at the same time? Let us first ask the analogous question within the context of the model introduced to illustrate Teichert's theory of correlation. Suppose we were asked to *explain why* two machines struck with the same note at the same time and in so doing left the same mark on a tape. We might reply that the machines were so constructed and so regulated that after having been put in place they perform in such a way as to produce a note of the same tone at the same time. If the machines had identical mechanisms and were closed to a high degree of approximation or, alternatively, were subject to the same perturbing influences, then it might be plausible to make such a claim. It is analogous to the claim that two clocks remain synchronized after having been separated from one another. We might, on the other hand, suggest that the machines are causally connected to each other; that the striking of a note in one machine *causes* the striking of a note of the same tone in another machine. If the machines are to strike at the same time, then the causal chain connecting them must be transmitted instantaneously.

Teichert's theory of correlation requires that biological communities in different places be "synchronized," or that biological signals be transmitted instantaneously, or both. Biological theory clearly forbids both of these assumptions. The problem goes beyond regarding something as a clock when it isn't. Teichert's theory is seriously defective as an account of paleontological correlation. It is a remarkable fact that Teichert knows this very well. In the very paper which contains the passages quoted above he explicitly rejects each of the assumptions which would lend credence to his theory. An examination of Teichert's paleontological work shows, furthermore, that he does not presuppose instantaneous signal velocities and synchronized biological events in his correlations. Teichert and the great number of other paleontologists who subscribe to this theory are not doing a bad job of correlating. They have, however, gone seriously wrong in their attempt to justify theoretically what they are doing. To accept this justification is to obscure thoroughly the only sound theoretical grounds upon which correlation can rest.

If a paleontologist were to encounter two sedimentary beds containing identical fossil assemblages, he might conclude that the represented faunas had come to each of the two places from a third place, or, putting it somewhat differently, that the two places were indirectly linked by a signal which had originated at a third place. A geologist would, analogously, link two mineralogically identical deposits of volcanic ash by hypothesizing indirect signal connection through an explosive volcanic eruption. Neither of these inferences is forbidden by any general or special theoretical considerations. If it were possible to determine the location C of the volcanic eruption that had caused the deposition of two beds of volcanic ash at A and B, the deposition events could be temporally ordered with respect to one another provided that the velocity of ash transport could be determined. If it turned out that the transit time between C and B was the same as the transit time between C and A, then it could be inferred that the deposition events were simultaneous. But we must stipulate more precisely the meaning of "simultaneous" here. Reichenbach states (1958, p. 128), "We can give a conceptual definition of 'simultaneity': two events at distant places are simultaneous if the time scales at the respective places indicate the same time value for these events." Suppose that our knowledge of the velocity of ash transport were precise enough to permit us to determine the number of days between the volcanic eruption at C and the depositions at A and B. If it turned out that the ash had arrived at A and B on the same day, the shortest interval of time measurable in these circumstances, then the time scales at the two places would, in effect, indicate the same time value for the events and it could be said that they were simultaneous. If arrival time could be determined more or less precisely than within a day, then the time scale and the meaning of simultaneity would be appropriately altered. Stratigraphers often invoke something very like Reichenbach's conceptual definition of simultaneity. Donovan, for example, says (1966, p. 32),

> Now, the mean duration of a Jurassic ammonite zone was about a million years, and most other zones were longer. It is clear that the time needed for wide dispersal is negligible compared with the duration of a zone, and the first appearance of a new species over a wide area can be regarded as simultaneous provided that the effects of facies can be discounted.

But any really meaningful application of the concept of simultaneity requires "time scales" which might be obtained through the measurement of signal velocities.

The measurement of signal velocities involves intriguing problems. Grünbaum (1963, p. 344) states:

It might be thought that distant simultaneity can be readily grounded on local simultaneity as follows. Suppose that two spatially separated events $E_1$ and $E_2$ produce effects which intersect at a sentient observer so as to produce the experience of sensed (intuitive) simultaneity at himself. Then this local simultaneity of the effects of $E_1$ and $E_2$ permits us to infer that the distant events occurred simultaneously, if the influence chains that emanated from them had appropriate *one-way velocities* as they traversed their respective distances to the location of the sentient observer. But this procedure is unavailing for the purpose of first characterizing the conditions under which separated events $E_1$ and $E_2$ can be held to be simultaneous. For the *one-way* velocities invoked by this procedure presuppose *one-way transit* times which are furnished by *synchronized* clocks at the locations of $E_1$ and $E_2$ respectively. And the conditions for synchronism of two spatially separated clocks, in turn, presuppose a criterion for the distant simultaneity of the events at these clocks.

But it is not necessary to have prior knowledge of simultaneity to measure certain kinds of signal velocity. A light signal may be transmitted from a point A to a point B and reflected back to A. If the distance between A and B and the elapsed one-clock time for the round trip are known, then the average round-trip velocity can easily be calculated. In order to base a determination of the one-way velocity of light upon the measured round-trip velocity, it is assumed that the time required to travel from A to B is the same as the time required to travel from B to A. Einstein pointed out that the concept of the simultaneity of distant events depended critically upon this assumption. He stated (1920, p. 4), "We have hitherto an A-time, and a B-time, but no common time to A and B. This last time (i.e., common time) can be defined, if we establish by definition that the time which light requires in traveling from A to B is equivalent to the time which light requires in traveling from B to A."

There is no standard signal in geology whose round-trip velocity has been measured and whose one-way velocity has been inferred on the basis of the assumption of constant velocity. Geologic theory permits neither the assumption that the average velocity of any two historic-geologic signals was the same nor the assumption that the velocity of any signal was constant. Knowledge that two events in geologic history occurred simultaneously would have to rest upon a knowledge of the average velocity of *the particular signal* employed in the inference.

The average velocity of a signal cannot be determined without prior knowledge of the temporal relationship of the events connected by it. This knowledge could in principle be obtained from radiometric clocks. But if we had this knowledge to begin with we would already know the very things that the signal connection was supposed to bring out. And if we decided to go ahead and measure the signal velocity anyway, we would obtain an item of knowledge that could not be employed in any other correlation. Causal connection, either direct or indirect, cannot lead us to a knowledge of the simultaneity of geological events.

Those concerned about the problem of historic-geologic signal velocities attempt to place broad limits upon velocities on the plausible assumption that the measured velocity of a signal in the present is more or less similar to that of a historical signal of similar character. Eames, et al. (1962, p. 130), for example, state, in justifying the stratigraphic usefulness of the Globigerinaceae,

> The planktonic habit of the Globigerinaceae assures rapid faunal (and genetic) migration; the observed density of living populations, which are free to migrate, indicates rapid reproduction in this superfamily (as also observed in other foraminiferal groups) and easy genetic exchange. All these factors reduce impediments to faunal migration and genetic interchange to a minimum.

We must always bear in mind, however, that a hypothesis that two events are ordered in the relation "simultaneous with" cannot be confirmed, because the confirmation would rest upon the determination of the average velocity of a particular historic-geologic signal. Hypotheses involving "earlier than—later than" relations can on the other hand be confirmed, because the confirmation rests only upon the knowledge of the direction of signal transit; knowledge which can, in principle, be obtained and in some cases may even be available.

A worldwide event which occurred at an instant in time or during a very restricted interval of time and which had detectable consequences at widely separated points on the earth's surface would constitute, in effect, an instantaneously transmitted signal linking those points. The possibility that events of this kind have occurred cannot be precluded in principle. To avoid the presupposition of simultaneity, however, the grounds for supposing that such an event had occurred would necessarily rest upon the use of conventional signals of finite, but unknown, velocity. The assumption that orogenies have occurred simultaneously at widely separated places on the surface of the earth has been used as a basis for correlation. The underlying assumption seems to be that whatever the sufficient conditions for the occurrence of an orogeny may be, they tend to be widely extended in space and narrowly extended in time. Attempts to demonstrate the simultaneity of widely separated orogenies on the basis of paleontological signals has been largely unsuccessful. At the same time, the theoretical grounds for supposing that orogenic episodes were worldwide have become less and less tenable. The hypothesis that other types of worldwide events—for example, climatic changes—have occurred during relatively short intervals of time is plausible but nonetheless difficult to substantiate.

I am not suggesting that geologists ever thought they could claim that two events occurred at exactly the same time. Nor do I maintain that it is meaningless to say that two geologic events occurred at the same time. I should not claim, for example, that because we do not know the velocities of the signals involved we cannot meaningfully say that some volcanic ash at two different localities had been deposited at the same time. The time required for volcanic ash transport is so small compared to the span of geologic time that we may consider that ash is transported instantaneously. The preceding analysis of correlation is not directed at improving the practice of stratigraphy. It is directed rather at providing a sound theoretical basis in terms of which that practice may be fully justified. So long as paleontological and physical correlation are regarded as methods directed toward and in principle capable of achieving a knowledge of simultaneity, its theoretical justification is obscured. Correlation is not a bad means of establishing simultaneity. It is not a means of establishing simultaneity at all. It is a method which under favorable circumstances permits the unambiguous ordering of events in "earlier than—later than." In my effort to eschew the role of arbiter in matters of stratigraphic research I do not wish to hide my conviction that a hard look at the theoretical foundations of correlation will lead us to conclusions of practical significance.

## ROCKS AND TIME

A geologist would be very much surprised, upon attending a meeting of historians, to find the participants engaged in a discussion of the problem of how to develop a set of terms which would allow a distinction to be made between divisions of historic time and the documents on the basis of which those divisions had been formulated. Yet during the past twenty-five years stratigraphers have become involved in an analogous discussion which raises the question of the distinction between "time terms" and "rock terms." The discussion has, for the most part, centered about terms, yet it is clear that the points at issue cannot be dismissed, as some have tended to do, as "mere" questions of semantics.

The pressing need for separate sets of terms for rocks and time intervals grew out of a conceptual confusion which, despite the heroic efforts of many stratigraphers, persists. It is remarkable that a distinction which, once it has been made, seems so natural and necessary should result in so much difficulty. The vigorous discussion, begun about twenty-five years ago, was precipitated by the recognition that so intimately entwined had become the concept of divisions of geologic time with the objects which constituted the basis for their inference, that it could indeed be said that no distinction whatever was being made.

The confusion between rocks and time intervals followed from the assumption that the physically defined boundaries of bodies of rock represented "isochronous surfaces," and that therefore bodies of rock somehow "correspond with" or were "equivalent to" certain time intervals. This assumption is clearly unwarranted on the basis of geologic theory. It makes as much sense to say that the western boundary of the Great Wall of China is isochronous as it does to say that the lower boundary of the Woodford formation is isochronous.

Opponents of the view that the boundaries of rock units are isochronous sometimes employ a vocabulary which perpetuates the very confusion between object and time which they seek to avoid. Is not the confusion evident, for example, in the statement, "A rock-stratigraphic unit may possess approximately isochronous boundaries, or its boundaries may transgress time horizons?" (From the Code of Stratigraphic Nomenclature, 1961, p. 649.) What does it mean to say that the boundaries of a rock-stratigraphic unit transgress time? One who was unaware of the conceptual and historical context in which this statement was made might understandably conclude that the statement meant no more than that rock boundaries persist in time. A geologist knows that it is not meant to say that the boundary which we observe in the present transgressed a time horizon, but rather that some events which account for the presence of the boundary transgressed a time horizon. To say that a boundary transgressed a time horizon, when what is meant is that events transgressed a time horizon, is confusing. When a geologist says, "The boundaries of a rock-stratigraphic unit may transgress a time horizon" he only means, "The event which marked the origin of a point (or limited area) on the surface of a rock-stratigraphic boundary may not have been simultaneous with the events marking the origin of every other point (or limited area) on the same boundary." Compare this statement to "The event which marked the origin of a point (or limited area) on the surface of the Great Wall of China may not have been simultaneous with the events marking the origin of every other point (or limited area) on the surface of the wall." Neither of these statements is surprising or particularly illuminating except to one who might, for some reason, have believed otherwise. This explication, incidentally, makes no recourse to the notion of a "time horizon."

EVENTS AND TIME

The distinction between rocks and the intervals of geologic time, however useful and illuminating it may be, does not by itself remove all of the conceptual difficulties that surround the notion of geologic time. It is true that rocks constitute the ultimate basis for the derivation of intervals of geologic time, but it is important to realize that the inferential gap between rocks and time intervals is a very wide one indeed. As a matter of fact, rocks on the one hand and intervals of geologic time on the other are at opposite ends of an inferential spectrum. Between these ends lies nearly all of our knowledge of the earth.

Historical inferences in geology lead to statements about events ordered in space and in time. Events are the "stuff" out of which our systematic knowledge of the earth is woven. Consequently the basis for any notions we may have about the divisions of geologic time must lie in a knowledge of events. The utility of the distinction between rock terms and time terms is that it may serve to remind us that intervals of time must be defined not by rocks but by events. The code tells us (p. 659), "The period comprised an interval of time defined by the beginning and ending of the deposition of a system." Thus the Cambrian period began with the first depositional event of the Cambrian and ended with the last depositional event of the Cambrian as, analogously, the administration of Herbert Hoover began with his inauguration and ended with the inauguration of Franklin D. Roosevelt. Now these seemingly plausible means of defining intervals may lead to difficulties, as the code points out in the following passage (pp. 659-60).

> To define periods rigorously in this manner is to create unnamed time units between periods, in other words, gaps in formal geologic time. By later work supplementary sections largely or wholly filling the hiatuses have been found elsewhere in the world and their rocks, by common consent, have been assigned to one or another of the contiguous systems. Many of the gaps have thereby been essentially filled. Today it is probable that formal geologic time as referred to actual rocks is continuous or even (as now classified) duplicated.

The analogy between geologic periods and presidential administrations is apparently not a very good one, because we do not tolerate the possibility of time intervals during two administrations at the same time, or intervals of time during no administration at all. Our initial reaction to this might be to say, "Of course it's a poor analogy. We know so much about the transition from one administration to another and so little about the transition from one geologic period to another." But a moment's reflection will show that the fact that there are no overlaps of, or gaps between, presidential administrations does not depend upon empirical knowledge at all, but rather upon the simple *device* of marking the end of one administration and the beginning of the next by the *same event.* Defining a period by the "beginning and ending of the deposition of a system" amounts to defining it by two events both of which are included in it. The period, so defined, constitutes a set of events closed at both ends. If an event marking the end of a period is not simultaneous with an event marking the beginning of the next period in succession, then a gap between or an overlap of the periods must result. The way out of this truly scandalous state of affairs is very simple. If the set of events defining a period is left open at one end, then the problem of gaps and duplications simply does not arise.

Suppose that presidential administrations began with an inauguration and ended with a closing ceremony, in which case administrations, like the code's periods, would constitute sets of events closed at both ends. We would have no gaps between administrations, or overlapping administrations, if we made sure that closing ceremonies and inaugurations took place at the same time. I am told that I quibble unnecessarily when I point out that it is impossible to say of any two geological events that they occurred simultaneously; that geologists can determine that some events occurred so close together in time that we may as well say they occurred at the same time. Yet the difficulty alluded to in the code would never have arisen if geologists could determine that two events occurred simultaneously, or if they thought they could, or even if, knowing that they couldn't, they thought they could come close enough "for all practical purposes." The distinguished *practicing stratigraphers* who wrote the code have encountered a real problem which is clearly rooted in their inability to say of two geologic events that they occurred at the same time. Their commitment to simultaneity and to the derivative notions of time planes and isochronous surfaces is, however, very deep. They say, for example (p. 658):

> Ideally the boundaries of time stratigraphic units, as extended geographically from the type section, are isochronous surfaces, representing everywhere the same horizon in time; thus ideally these boundaries are independent of lithology, fossil content, or any other material basis of stratigraphic division. In actual practice, the geographic extension of a time-stratigraphic unit is influenced and generally controlled by stratigraphic features.

Although I have admitted that some geologic events occured so close together in time that they were for all practical purposes simultaneous, I am by no means willing to grant that we can meaningfully talk about the worldwide extension of isochronous surfaces or time planes which, I take it, would require for their definition a set of at least three simultaneous events. Biological signals are almost always employed in the broad extension of time planes. There is every reason to suppose that biological signals have, on the whole, relatively low velocities, that the velocities vary greatly from signal to signal, and that knowledge of signal direction is extremely difficult to come by. These characteristics, together with the great distances involved in intercontinental correlations, conspire to undermine our confidence in the extension of time planes on a worldwide scale.

Why, in the face of these very substantial difficulties, are stratigraphers so preoccupied with the establishment of time horizons? Why do they, in fact, seem to believe that the spatial extension of temporal ordering *requires* the use of time planes? Suppose that we were to ask whether or not an event that had occurred in North America was a Cambrian event. If the beginning and ending of the Cambrian were marked by a time horizon, then we will have recognized events in North America which occurred at the same time, or at nearly the same time as the events which marked the beginning and ending of the Cambrian at the type locality. Whether or not the event in question is or is not a Cambrian event depends upon whether it did or did not occur during the interval defined by the first and last Cambrian events or, according to my system, during the interval defined by the beginner event of the Cambrian and the beginner event of the Ordovician. This may be determined if the event can be linked with a Cambrian event at the type locality by a causal chain which may or may not include the event which is said to define the time horizon in North America. Alternatively, the event in question may be ordered by superposition with respect to events which are causally linked to events in the type area by a chain which may or may not include the event which is said to mark the time plane in North America. To infer that a sedimentary bed was deposited during the Cambrian does not depend upon the designation of a time plane any more than to infer that a man was born during the Renaissance depends upon the designation of a set of events simultaneous with, or nearly simultaneous with, the death of Dante. We can in principle and in fact decide within which interval of time an event occurred without any reference whatever to a time plane or isochronous surface. How are we to understand the role of the time plane in stratigraphic inference, particularly in view of the fact that this role apparently does *not* depend upon the events defining a time plane being *demonstrably* simultaneous, but only upon our saying that they are simultaneous?

I do not propose that we purge our minds of the notion of simultaneity. It is an important item of our conceptual baggage, both in everyday life and in geology. I do propose that we abandon simultaneity and the time plane of worldwide extension as an essential feature of our *stratigraphic system*. A formal interval of geologic time may be defined by a beginner event included in it and a beginner event included in the next interval in temporal succession. This definition in no way depends upon the knowledge, or assumption, or hypothesis that a beginner event is simultaneous with any other event. It is a peculiar consequence of selecting bodies of rock as the basis for defining intervals of geologic time that the events which begin intervals of time must be in the *same place* as the events which end them. Historians have no such requirement. The Middle Ages may begin in Athens with the closing of the Academy by Justinian and may end in Umbria with the death of Nicolas of Cusa. Thus there is no need to infer simultaneity in order to avoid gaps and overlaps between periods.

## References Cited

AMERICAN COMMISSION ON STRATIGRAPHIC NOMENCLATURE, 1961, Code of stratigraphic nomenclature: Am. Assoc. Petroleum Geologists Bull., v. 45, p. 645-665.

DONOVAN, D. T., 1966, Stratigraphy: an introduction to principles: London, Murby, 199 p.

EAMES, F. E., BANNER, F. T., BLOW, W. H., and CLARKE, W. J., 1962, Fundamentals of mid-Tertiary stratigraphical correlation: Cambridge, Cambridge University Press, vii+163 p.

EINSTEIN, A., 1920, On the electrodynamics of moving bodies, *in* The principle of relativity, Original papers by A. Einstein and D. Minkowski, translated into English by M. N. Saha and S. N. Bose: Calcutta, University of Calcutta, xxiii+186 p.

GRUNBAUM, A., 1963, Philosophical problems of space and time: New York, Alfred A. Knopf, xi+448 p.

———— 1964, The anisotrophy of time: Monist, v. 48, p. 217-247.

JELETSKY, J. A., 1956, Paleontology, basis of practical geochronology: Am. Assoc. Petroleum Geologists Bull., v. 40, p. 679-706.

NEWTON, I., 1947, Principia, ed. F. Cajori: Berkeley, University of California Press.

REICHENBACH, H., 1958, The philosophy of space and time: New York, Dover Publications, xvi+295 p. (English trans. of Philosophie der Raum-Zeit-Lehre published in 1928).

TEICHERT, C., 1958, Some biostratigraphic concepts: Geol. Soc. America Bull., v. 69, p. 99-120.

WHITROW, G. J., 1961, The natural philosophy of time: London, Thomas Nelson & Sons, xi+324 p.

CHAPTER EIGHT

# Paleontology and Evolutionary Theory

IN THE INTRODUCTION to the collection *Models in Paleobiology*, Schopf (1972, p. 4) says:

> We look with awe at the results which have followed the application of fundamental chemistry in a geochemical context, and the use of cartesian methodology in molecular biology. Our fossils are amenable to more than the traditional methods of analysis. We wanted to write a book which would take as its goal the self-conscious use of models in paleontological research.

This kind of lament may have a familiar ring, particularly for those paleontologists who lived through the movement of the 1950s in which the "new systematics" was applied to paleontology. There were then and there are now paleontologists who wish to escape "pure description" and establish paleontology once and for all on an equal footing with such disciplines as geochemistry and molecular biology. The accomplishment of this task depends upon an understanding of the relationship between paleontology and biological theory. It is this critical relationship that I shall examine in the following pages.

## FOSSILS

Paleontologists often claim that fossils tell us something. But fossils, by themselves, tell us nothing; not even that they are fossils. Certain objects found entombed in rocks do not point intrinsically beyond themselves to organisms. When a paleontologist decides whether or not something found in the rock is the *remains* of an organism he decides in effect, whether or not it is necessary to invoke a biological event in the explanation of that thing. Sometimes a judgment is made without hesitation because the object in question resembles some part of a living organism. Steno argued that certain objects he found enclosed in rock were the remains of an animal because of their striking resemblance to the teeth of living sharks. But paleontologists encounter objects which do not resemble the parts of any living plant or animal and yet conclude that they are fossils. There is no structural feature which serves to distinguish fossils from entities of every other kind. A paleontologist attempts to account for a problematic object without assuming some antecedent biological event. After repeated failure to explain an object on physical grounds alone it may come to be regarded as a fossil. An object ceases to be regarded as a fossil when a physical explanation of its origin is accepted. The "jellyfish" of the Nankoweap group are rejected when Cloud's (1960, p. 43) claim that they are gas bubble markings is accepted.

Fossils, literally and figuratively lifeless, require the paleontologist to assume very little about the character of the organisms whose remains they are supposed to be. The properties of the plants and animals of history are not given to us in immediate experience. They must be reached within inferences that invoke biological theories. When a paleontologist concludes that he is dealing with a fossil and by implication with an organism, he already knows a great deal about that organism. He knows that the plants and animals which he encountered in the geologic past will have the characteristics required of them by the theory which he has presupposed in order to reach them. A paleontologist will not turn to the fossil record in search of answers to questions that are already answered within the biological theory to which he subscribes. He goes to the fossil record with questions directed at the ways in which the presupposed theory permits organisms to differ. He will not ask whether or not an organism metabolized, but he may ask whether or not it ate berries. All this is by way of saying that paleontology is regarded as a historical, rather than theoretical, discipline by most of those who practice it. There is, moreover, a near consensus among paleontologists as to what constitutes the unproblematic biological theory to be invoked in historical inferences. What theoretical disputes do arise are likely to involve evolutionary theory, apparently because this theory, unlike others such as biochemistry and embryology, is supposed to be in some unique way *historical.*

## FOSSILS AND EVOLUTIONARY THEORY

### *A*

Goudge (1961) expresses the view that evolutionary theory is uniquely historical in the following passage (pp. 16-17):

> The theory of evolution, like any comprehensive scientific doctrine (e.g., modern atomic theory), has a powerful unifying function. It brings order

into a large array of empirical evidence and helps to integrate the findings of numerous special disciplines. Like modern atomic theory it is abstract in character and requires for its formulation concepts which cannot be correlated with what is directly observable. In other respects, however, evolutionary theory is radically unlike comprehensive theories in physics and chemistry. For the latter are wholly systematic and non-historical, whereas the former combines systematic and historical elements. This "double-barrelled" feature of evolutionary theory is one of its most distinctive characteristics. Part of the theory utilizes historical data in the form of fossils, reconstructs unique, non-recurrent evolutionary histories, and advances historical explanations. Another part of the theory utilizes non-historical data derived from observation of contemporary organisms, incorporates many of the generalizations of such special sciences as genetics, ecology, embryology, comparative anatomy, etc., and makes use of law-like statements in formulating systematic explanations. The logical relations within each of these parts and between them are by no means easy to disentangle.

Just what does this double-barrelled feature of evolutionary theory amount to? In an attempt to answer this question let us examine an evolutionary theory. Darwin said in the summary of Chapter IV of *The Origin of Species* (pp. 102-3):

> If under changing conditions of life organic beings present individual differences in almost every part of their structure, and this cannot be disputed; if there be, owing to their geometrical rate of increase, a severe struggle for life at some age, season, or year, and this certainly cannot be disputed; then, considering the infinite complexity of the relations of all organic beings to each other and to their conditions of life, causing an infinite diversity in structure, constitution, and habits, to be advantageous to them, it would be a most extraordinary fact if no variations had ever occurred useful to each being's own welfare, in the same manner as so many variations have occurred useful to man. But if variations useful to any organic being ever do occur, assuredly individuals thus characterized will have the best chance of being preserved in the struggle for life; and from the strong principle of inheritance, these will tend to produce offspring similarly characterized. This principle of preservation, or survival of the fittest, I have called Natural Selection. It leads to the improvement of each creature in relation to its organic and inorganic conditions of life; and consequently, in most cases, to what must be regarded as an advance in organization.

Darwin makes a claim that is common in scientific discourse. If certain fundamental propositions are granted, he asserts, then other propositions necessarily follow. The propositions which, if granted, lead to the conclusion that evolution has occurred are, according to Darwin, the following:

1. Under changing conditions of life organic beings present individual differences in almost every part of their structure.

2. The size of populations increases at a geometrical rate.

3. There is an infinite complexity of the relations of all organic beings to each other and to their conditions of life.

4. There is a strong principle of inheritance.

These propositions attribute certain properties and dispositions to organisms and to populations of organisms in much the way the laws of mechanics attribute certain properties and dispositions to physical objects and to groups of physical objects. Now the claim is made by Darwin that the principles of his theory have *historical consequences*, which is to say that they may be used in the construction of unique, nonrecurrent evolutionary histories and in the advancement of historical explanations. Grene (1961, p. 42) has gone so far as to say that "evolutionary theory is essentially an assessment of the past." An assessment of the past must, I take it, include more than a statement of general principles. Grene has apparently included in evolutionary theory not only its general principles but also the singular descriptive statements about the past that may be inferred by invoking these principles. But, of course, any body of general propositions may be employed as an instrument of historical inference. If we were to include within "classical mechanics" the descriptive historical statements that may be inferred on the basis of its laws, then it too might be regarded as an assessment of the past. Classical mechanics is used by geologists in the reconstruction of unique, nonrecurrent histories and in the advancement of historical explanations, but it is not for this reason considered to be "double-barreled."

What has been said about the character of Darwinian theory could be said as well about Lamarckian and synthetic theory. All of them attribute properties and dispositions to organisms which are said to result in certain historical consequences. There is no *unique* and mysterious historical dimension in any of these theories. They may be employed as instruments of historical inference in just the way that physical and other biological theories are so employed. But the claim of special historical status for evolutionary theory is not completely empty, if only because those who use it are usually, although by no means always, concerned with its historical consequences. And more interesting and significant from a methodological point of view is the fact that evolutionists, unlike physical scientists, routinely present descriptions of historical events as instances of confirmation and falsification for their theories. Let us now consider this practice.

*B*

A paleontologist who subscribes to synthetic evolutionary theory will attribute to organisms past, present, and future certain fundamental characteristics required by that theory. Difficulties arise because evolutionary theorists are not in complete agreement as to what the theory requires of biological events; and furthermore, in their effort to achieve agreement they may explicate the concepts of the theory in such a way that their relevance to paleontology is obscured or even lost.

I shall limit the discussion to a consideration of some difficulties that have arisen in connection with the term *adaptation*. The concept *adaptation* is a particularly interesting one because the claim is sometimes made that the determination of whether or not fossil organisms were adapted is critical for assessing the validity of synthetic theory.

*Adaptation* has been used to designate both *a state* and *a process by which a state is achieved*. When used in the latter sense it is almost always connected with one or another historical, or evolutionary, system of explanation. I shall use the term *adaptation* only to designate a state. Because the question of whether or not an entity is adapted is frequently cited as having relevance to the validity of some evolutionary theory, it is essential that the concept not presuppose the theory to whose validity it is supposed to be relevant.

*Adaptation* is frequently applied to a part, or an aspect, or an activity of an organism. It is in this sense that it is most commonly employed by paleontologists. Simpson says, for example (1953, p. 160), "For our purposes, *an* adaptation is a characteristic of an organism advantageous to it or to the conspecific group in which it lives." Grant (1963, p. 94) uses the term *adaptation* in this conventional way when he says, "The webbed feet of a duck set toward the rear of the body represent an adaptation for swimming." To say that webbed feet are an adaptation, or to say that they are adapted for swimming, is to say that they are disposed, under certain circumstances and at certain times, to function in such a way as to aid in swimming.

The functions of many parts seem to be clearly evident. It takes no great insight to ascribe a paddling function to the feet of a duck, but one need only consider the history of attempts to understand the function of the vertebrate heart to realize that ascribing a function to a part sometimes involves a complex series of hypotheses and tests. The most direct test procedure is to remove, or otherwise render useless, the part in question and then to determine if the function attributed to it is still performed. Difficulties arise in connection with tests of this kind mainly because an organism is supposed to be an integrated whole in which the designated parts have activities which are complexly related to the activities of other parts. Even when a biologist satisfies himself as to the activity of a part, it is not always a simple matter to show that the activity is advantageous to the organism.

Despite these and other difficulties, functional statements are considered meaningful by biologists and tests are accepted in support of them. Harris's (1936) studies on the function of dogfish fins may be regarded as an attempt to demonstrate the adaptedness of parts. The activity of the pectoral fins, a part of the dogfish, under environmental circumstances and during a time interval which may be specified, have the effect of maintaining pitch equilibrium. The maintenance of pitch equilibrium is advantageous to the dogfish in the specified environment and during the specified interval of time. These statements are supported by dramatic experiments involving the removal of fins.

*C*

Paleontologists can take little comfort from the confidence with which neontologists make functional statements. They cannot observe the activity of organisms they study at all, let alone observe it under controlled conditions. Paleontologists do, however, attribute functions to the parts of fossil organisms by a principle of analogy. They maintain, in effect, that if a part of a fossil organism is similar to a part of a living, closely related organism, then it had a similar or identical function. Beerbower (1968, p. 479) says, for example:

> It seems reasonable to suggest that the carnivorous therapsids, with secondary palate and well developed chewing mechanisms (teeth and muscles), had an internally controlled body temperature—were *endotherms.* This is further confirmed by the development of bony scrolls in the nasal air passages, since, in recent mammals, these bear mucous membranes that warm and filter the incoming air.

But when Romer (1966, p. 176 and elsewhere) and others suggest that the "sail" of certain pelycosaurs may have functioned as a temperature regulating device, they do not liken it to a structure of some living organism. The hypothesis is no less plausible for this. Rudwick (1964) has recently shed a good deal of light on the method by which activity and function are inferred from part. The recognition of a structural similarity between a part of a living organism and a part of fossil organism does not, according to Rudwick, establish that the parts have the same function. The similarity may only suggest a hypothesis concerning the function of the part of the fossil organism. The hypothesis must be tested by asking certain critical questions and devising a means of answering them. Rudwick says (pp. 35-36):

> How can this function possibly be fulfilled? What operational principles are involved? What basic characteristics must any such mechanism possess?

What, in the most general terms, is the specification to which any structure must conform, if it is to fulfill this function effectively? Such questions lead us to a *generalised* structural specification for the function (for example, a general specification for aerofoils for gliding). But this will probably be too vague and imprecise to be of any great use. The specification can be made more rigorous by taking into account the limitations imposed by the properties of the material in which it would have to be embodied—if at all—in the fossil structure (for example, in a pterodactyl, the anatomical relation between the forelimb skeleton and the wing-membrane, and the finite strength of these materials, would presumably limit the possible range of "design" of the aerofoil). Such a specification then describes *the structure that would be capable of fulfilling the function with the maximal efficiency attainable under the limitations imposed by the nature of the materials.*

Rudwick is at pains to point out (p. 30) that the method which he proposes cannot lead to a conclusion that a structure did *in fact* fulfill a certain function, but only that it would have been physically capable of fulfilling that function. There is thus a fundamental ambiguity involved in the attempt to ascribe a function to a part of a fossil organism. We must be careful, however, not to attach a special methodological significance to this fact. Of *any* object whose activity has not been observed it can only be said that it *could have* fulfilled some function. It can never be said that it *did in fact* fulfill that function. This principle is as applicable to artifacts as it is to parts of organisms.

*D*

Paleontologists often speak of a population, or species, or higher category as being adapted. Jepsen says, for example (1949, p. 488),

> Multituberculates were very successful for a very long time. Their known history ranges from Late Jurassic to Early Eocene, a period of approximately seventy-five million years. Only two other orders of mammals, the marsupials and the insectivores, with a phylogeny beginning in the Late Cretaceous, have equally long histories. This durability of the Multituberculata and the fact that the order contains more genera than most orders of living mammals suggest a high degree of adaptation and adaptive plasticity.

The statement (p. 489), "They [the multituberculates] were abundant during the Paleocene not only as generic groups but as individuals as well," suggests that numbers of individuals is another criterion for success. How long must a group of organisms persist, and what degree of diversity must be achieved within the group, and how abundant must its members become, before it is considered to be successful? The briefest consideration of the use of the term "success" by paleontologists reveals that it cannot simply be said that a group is successful but rather that it is more or less successful than some other group. Thus Simpson says (1953, p. 161), "If one group is more successful than another under given (the same or similar) conditions of life, it is fair to conclude that it is better adapted. The surest criteria of such success is increase in relative abundance of the better adapted group."

But serious difficulties arise in connection with this notion as Lewontin (1957, pp. 395-96) points out, ". . . in nature no two populations can exist in identical environments because the genotypic composition of a population defines, in part, the biotic environment and because no two populations have identical genotypic compositions."

The difficulties that arise out of attempts to explicate "adapted population" in terms of relative success are mitigated by the fact that being "successful" is not regarded as the same thing as being "adapted." Success only suggests adaptation. It does not explain it. Simpson, in commenting upon Jepsen's (1949) suggestion that the decline of the multituberculates and the rise of the rodents may be an example of competitive replacement, remarks (1953, p. 300):

> The two groups are markedly different in ancestry and many features of anatomy, but strikingly similar in rodent like adaptation. The decline of the other group as rodents appeared and began to expand is evident in the table. Moreover, multituberculates are relatively abundant in all known late Paleocene local faunas but one (Bear Creek, Montana), where they are absent; precisely there is the only known occurrence of Paleocene rodents.
>
> In such cases a complete explanation would also state why the succeeding competitor was superior. This can seldom be clearly stated, but some suggestion is often possible. In the example just given, Jepsen points out that rodent incisors are clearly more efficient mechanically than those of multituberculates.

A complete explanation of the competitive superiority of the rodents, Simpson maintains, requires ultimate reference to parts and their functions. The function of parts is central to a paleontologist's notion of success and adaptedness. This is clearly stated by Schaeffer in his discussion of broad adaptation and biological improvement. He says (1965, p. 319):

> The most convincing evidence for biological improvement through time can be found in vertebrates, which have numerous mobile skeletal units associated with both feeding and locomotion. Such evidence must be obtained from the fossil record and from consideration of the closest living relatives or counterparts of the extinct forms under consideration. The paleontological record can be meaningfully utilized only when it is possible to make reasonable inferences regarding the function of the skeletal units and the surrounding soft tissue structures.

*E*

Rudwick (1964, p. 27) attributes to synthetic theorists the claim that the adaptation of features is universal. This claim must, according to Rudwick, be tested against the fossil record. He states (1964, pp. 39-40):

With a certain critical fossil organism before us, we may argue whether its distinctive features were or were not adaptive. But we cannot reach any conclusion about these features except by testing them as possible embodiments or conceivable operational principles. We may then be able to demonstrate that they were probably adaptations of high efficiency for particular functions; and by doing so we shall add cumulative weight to the Synthetic theory (or any other theory that stresses the ubiquity of genuine adaptation). Yet if we are unable to do so, our failure will not add corresponding weight to some other theory, for it can always be said that the features may prove to have been adaptive, if only we can think of the right function. For there is no positive criterion by which nonadaptedness can be recognized and demonstrated. At least in this respect, therefore, the Synthetic theory would seem to be as unfalsifiable as its rivals are unverifiable.

The fact that there is no positive criterion by which nonadaptedness may be recognized is not related to the functional character of the concept. To clarify this point, let us suppose that a geologist were induced to seek in a sequence of historical events an instance of confirmation for Newton's third law that *to every action there is an equal and opposite reaction*, a statement of the same form as the one attributing adaptation to all parts of organisms. The geologist would immediately find that he was engaged in a very unfamiliar kind of activity. Ordinarily if a geologist has doubts about the validity of some physical law he will turn to a physics book. But consider the following hypothetical case: A geologist finds a volcanic bomb resting upon a smooth surface of basalt. Beneath the bomb is a set of scratches and cracks which he concludes were formed when the bomb struck the surface. A careful examination of all of the relevant evidence permits the conclusion that the *action* of the bomb striking the basalt surface had been met by an opposite reaction. This inferred succession of events supports the third law in exactly the way that showing a feature to be a possible embodiment of some conceivable operational principles supports an evolutionary theory that stresses the ubiquity of adaptation. Is there any positive criterion by which nonreaction can be recognized? Certainly finding a basalt bomb resting upon a surface with no evidence of a point of impact provides no such criterion. A physicist would simply deny that such a case had anything to do with the verification of the third law. He might say, "If we knew all of the relevant circumstances then the law would be supported."

Is Newton's third law unfalsifiable? In Rudwick's peculiar sense of "unfalsifiable" it is. For him a universal proposition is unfalsifiable if we can ever point to a state but cannot, on independent grounds, point to another state with which the universal proposition requires it to be linked. We cannot show that every feature of every organism which we encounter is adapted. We cannot show that every action force we encounter is opposed by a reaction force either, but we do not call attention to this fact, nor do we say that the third law is consequently unfalsifiable.

Unfalsifiability, in the sense that Rudwick uses it, is not a curious methodological dilemma which plagues evolutionary investigations, nor is it a devilish device which permits synthetic theorists to operate beyond the restrictive canons that define legitimate scientific discourse. It is rather the inevitable outcome of the requirement that a law be tested under circumstances which permit a certain minimum degree of control over, or knowledge of, the initial and boundary conditions which, according to the law to be tested, are relevant. For the precise reason that in historical situations we almost never have such control and knowledge, we do not consider the tests of laws and theories to be "necessarily in part historical." History permits scientists to accept evidence which supports a law and to reject evidence which does not. Anyone who subscribes to Newton's second law must be prepared to point to tests that can be performed here and now under controlled conditions which may count as instances of confirmation for it. Anyone who subscribes to the proposition "every feature of every organism is adapted" must be prepared to do the same thing. The proposition must be judged on the basis of these tests.

*F*

In an attempt to make a point about the relationship between the law statements of a scientific theory and historical events, I have up to now ignored the fact that most exponents of synthetic theory have denied that the adaptation of features or parts is universal. Genetic drift, moreover, plays little part in this denial. A more distinguished spokesman for synthetic theory than Dobzhansky could not be found. He says (1956, p. 339), "We need not follow the extremists who insist that all evolutionary changes are adaptive; however, we have to suppose that most organs and functions of most organisms are, or at least were at the time when they were formed, in some way useful to their possessors." Simpson has repeatedly offered his opinion that nonadapted characters are permitted by synthetic theory (see, for example, Simpson, 1944 and 1953). These evolutionists do not expect to find that every part of every organism is adapted. The question of the adaptedness of parts is for them a historical question. If they were to encounter a part which did not seem to be adapted they might say, "If we knew all the relevant circumstances, and if we could only think of the right function, then it would become clear that the part is adapted." They might, on the other hand, say, "This part is probably not adapted."

Paleontologists of Darwinian inclination approach their subject matter with the working hypothesis that, as Dobzhansky puts it, "most organs and functions of most organisms are, or at least were at the time when they were formed, in some way useful to their possessors." But this working hypothesis has little bearing on the validity of the theory. A proposition to the effect that *most* As are also Bs is not falsified by finding *an* A that is not a B. There may be a question as to whether the fundamental postulates of synthetic theory are consistent with the notion of nonadapted parts. But this is not the same question as "Are there nonadapted parts?" It cannot be answered by examining organisms, fossils or otherwise. It can only be answered by examining arguments.

*G*

There is a sense of "adapted" that is attributed universally, if not as a consequence of synthetic theory, by workers who are strongly inclined toward the synthetic position. Pittendrigh (1958, p. 394) states, ". . . to say that things are organized is to say that they are adapted." And (p. 395),

> By identifying organization and adaptation the latter concept is put in proper perspective. The study of adaptation is not an optional preoccupation with fascinating fragments of natural history; it is the core of biological study. The organism is not just a system some features of which may or may not be adaptive; the living system is all adaptation insofar as it is organized.

Pittendrigh leaves no doubt that he means to claim that all organisms are adapted. For him adaptation is not just a characteristic which all organisms happen to have. It is one of the fundamental characteristics which distinguish organisms from nonorganisms. He certainly believes that in order to *understand* organisms we must recognize that they are adapted. It would not be overstating the case to say that his theory of organisms *requires* that all organisms be adapted. What would Pittendrigh's reaction be if he were presented with a historical event that was purported to falsify his hypothesis that all organisms were adapted? He would most certainly claim, I think, that such an event could not occur, and that the inference which led to the event must be in some way mistaken. He would respond in just the way that a physicist would respond upon being told that a Cambrian event falsified the second law of thermodynamics.

A number of questions might be put to Pittendrigh regarding his concept of adaptation. What, for example, are some specific tests that could be presented in support of the assertion that this kind of adaptation is universal? This is to ask, "What are the empirical consequences of the hypothesis?" And (it is here that the matter of *falsifiability in principle* might significantly arise) is there any conceivable observation or set of observations that he would accept as a falsification of the hypothesis? Intimately related to these questions is that of the precise explanatory role which the hypothesis plays in synthetic theory and its relationship to other hypotheses containing the term *adaptation.* An examination of the fossil record will not provide answers to any of these questions.

*H*

We cannot fault paleontologists of whatever theoretical persuasion for refusing to accept events out of history as falsifications of the general laws to which they are committed. For a paleontologist to be committed to such a law means that he will assume, at least for the sake of historical inferences, that events which falsify it cannot occur. This is not a kind of dogmatism directed at shaping the world according to some arbitrary preconceived scheme. It is a means of getting from the present to the past.

If a theory permits us to suppose that an organism may or may not have some characteristic, then our judgment as to whether the organism does or does not have that characteristic is historical and has no bearing on the validity of the theory. If a theory requires that an organism have some property, then the judgment as to whether or not the organism does or does not have that property is of theoretical significance. But fossils do not present us with organisms which may serve as test events for biological theories. The animals and plants of the geologic past possess biological properties only by virtue of the theory presupposed to infer them. It is true, of course, that the testing of any theory requires that a set of assumptions which Lakatos (1970) has called "the unproblematic background" be made. Living organisms cannot serve as test events for a theory without such a system of presuppositions. But the number of assumptions and the complexity of the inference required to get at a fossil organism is very much greater than that required to get at a recent organism. To "flesh out" a fossil organism we almost always have to presuppose the properties of greatest theoretical interest. For this reason extant plants and animals will always take precedence over fossil organisms in the testing of biological theories.

## EVOLUTIONARY THEORY AND THE FOSSIL RECORD

### *A*

The fundamental propositions of evolutionary theory have no more historical dimension than do the axioms of classical mechanics and can therefore be tested against events in the present. The purported consequences of the propositions are often long-range historical sequences which cannot be tested directly. The obvious distinction between the properties of organisms and of groups of organisms and the historical consequences of attributing these properties is frequently not drawn, principally because many biologists have fallen into the habit of supposing that we can "see" evolution occurring. On this view our principal reason for supposing that evolution has occurred over long periods in the past is based upon an extrapolation of trends observable in our own lifetime. No wonder then that some are led to ask why evolution was not "discovered" long before the eighteenth century. A view that does much greater justice to the theoretical complexity of evolutionary studies is one in which what can be observed is regarded as providing empirical support for those principles which, according to evolutionary theory, entail the occurrence of evolution.

One of the great strengths of Darwinian theory has been that it purports not only to provide an explanation of organic evolution, but also to provide a promising means of testing the claim that evolution has occurred. For Darwin the evolution of organisms followed necessarily from certain premises which, he claimed, must be granted. Evolution, at least in the sense that Darwin speaks of it, cannot be detected within the lifetime of a single observer. Darwinian theory, however, is supposed to have, in addition to evolution, other less sweeping consequences which are more amenable to observational test. A great deal of investigation during the past century has been directed toward the accumulation of observational and experimental evidence that bears upon the validity of Darwinian theory. Every consequence of the theory that can be verified by more or less direct observation will be counted by Darwinians as indirect evidence in support of the contention that evolution has occurred.

### *B*

The claim is made that paleontology provides a *direct* way to get at the major events of organic history and that, furthermore, it provides a means of testing evolutionary theories. This claim raises the critical question of how close we can get to evolution without presupposing some causal theory of descent. With the assumption of the geological apparatus of temporal ordering (for a discussion of this subject, see pp. 128-47) the paleontologist arrives at a distribution of organisms in space and time. The organisms will have the properties imposed upon them by whatever biological principles have been presupposed in their inference. These properties need not include any required by a causal theory of evolution. The temporal and spatial distribution is not entailed by any biological theory, but by the ordering principles of geology. Thus the paleontologist can provide knowledge that cannot be provided by biological principles alone. But he cannot provide us with evolution. We can leave the fossil record free of *a theory of evolution*. An evolutionist, however, cannot leave the fossil record free of the *evolutionary hypothesis*.

But the danger of circularity is still present. For most biologists the strongest reason for accepting the evolutionary hypothesis is their acceptance of some theory that entails it. There is another difficulty. The temporal ordering of biological events beyond the local section may critically involve paleontological correlation, which necessarily presupposes the nonrepeatability of organic events in geologic history. There are various justifications for this assumption, but for almost all contemporary paleontologists it rests upon the acceptance of the evolutionary hypothesis. Despite these pitfalls we can with reasonable care avoid the danger of presupposing what it is we want ultimately to test, and have at our disposal a distribution of organisms in space and time that we suppose to have been related to one another by descent. Something more is, however, needed. When paleontologists invoke paleontological evidence in support of evolutionary theories, that evidence invariably includes assertions about the particular relationship of one fossil organism to another, which is to say, assertions about phylogeny. I have misgivings about the use of phylogenies as instruments of theoretical investigation, but they do not stem from the fact that phylogeny construction obviously presupposes whatever theoretical principles they purport to test. They are grounded rather in the belief that, despite some valiant and interesting efforts, paleontological phylogeny construction has not been provided with a solid theoretical foundation (for a review of the problems of phylogeny construction see Ghiselin, 1972). Providing this foundation is, in my opinion, the most urgent task now facing theoretically disposed paleontologists.[1]

*C*

Despite the bright promise that paleontology provides a means of "seeing" evolution, it has presented some nasty difficulties for evolutionists, the most notorious of which is the presence of "gaps" in the fossil record. Evolution requires intermediate forms between species, and paleontology does not provide them. The gaps must therefore be a contingent feature of the record. Darwin was concerned enough about this problem to devote a chapter of the *Origin* to it. He accounts for "the imperfections of the geological record" largely on the basis of the lack of continuous deposition of sediments and by erosion. Darwin also holds out the hope that some of the gaps would be filled as the result of subsequent collecting. But most of the gaps were still there a century later, and some paleontologists were no longer willing to explain them away geologically. Simpson was the most prominent among them. He said (1953, p. 361):

> It is thus still too soon for the rest of us to take the discontinuities of the paleontological record for granted. Even apart from that, the recognition and interpretation of such discontinuities is interesting and is a necessary, frequently also a practical and useful, part of the paleontological profession. Moreover, it is a fact that discontinuities are almost always and systematically present at the origin of really high categories, and, like any other systematic feature of the record, this requires explanation.

The "fact that discontinuities are almost always and systematically present at the origin of really high categories" is an item of genuinely historical knowledge because it rests necessarily upon historical inferences.

Simpson holds that speciation can be explained by attributing to organisms certain characteristics which are required by synthetic theory. He would hold further that experiments and observations could be presented to support more or less directly the conviction that organisms do, in fact, have these characteristics. Simpson, in effect, attempts to explain the kind of historical event "origin of higher categories" by attributing these same characteristics to the organisms of the past. He makes no claim that an examination of fossils directly supports his doing so. In connection with this point he says (1944, p. xvii), "One cannot directly study heredity in fossils, but one can assume that some, if not all, of its mechanisms were the same as those revealed by recent organisms in the laboratory."

It may be asked, "If it is assumed that the organisms of the past had the fundamental characteristics which synthetic theory attributes to all organisms, then can the class of historical event 'origin of higher categories' be explained?" Simpson believes that he has succeeded in accounting for the origin of higher categories in terms of the same "factors" which account for the origin of lower categories. He says (1953, p. 376):

> This general survey of higher categories shows that their evolution has special features tending to distinguish it from the evolution of lower categories. It may be in some cases but it is not typically merely a multiple of the evolution of species and other lower categories. The distinguishing features related mainly to the scale and the adaptive relationships of the evolution of higher categories. They involve certain durations, intensities, and combinations of factors. There is no reason to believe that any different factors are involved than those seen in lower categories or in "microevolution." On the contrary, those factors are fully consistent with what we know of higher category evolution and quite capable of explaining it.

Schindewolf (see particularly Schindewolf, 1950, pp. 380-431) does not agree that synthetic theory can account for features of history that Simpson calls "the evolution of higher categories." He claims, furthermore, that he can account for them by invoking a different set of principles. To put it this way, however, fails to do justice to the complexity of the disagreement between these two evolutionists, for it sounds as if they have agreed that some kinds of events have occurred and simply present alternative means of explaining them. There is, of course, agreement to a certain point. Simpson and Schindewolf would agree that the fossil record is relevant to evolution and would further agree as to the order of succession of various kinds of organisms through history. Either theory can account for these fundamental features of the record. But the discussion of the disagreement is not carried on at this level. Schindewolf claims that synthetic theory cannot account for the cataclysmic origin of new adaptive types. But Simpson does not feel called upon to explain this, since he believes that it does not occur. Nor does Schindewolf feel called upon to explain the origin of new adaptive types by a less abrupt and sequential acquisition of characters, since he believes that this does not occur. The historical events to be explained in each case presuppose the theories that are invoked to explain them. Schindewolf's theory simply does not permit new adaptive types to originate by the gradual acquisition of characters, and Simpson's theory simply does not permit new adaptive types to originate cataclysmically. The theoretical presuppositions of the two affect their view of historical events in a less direct way, for although they may agree about the order of succession of organic events, their theoretical notions lead them to different conclusions about the magnitude of the time intervals between events, and this in turn will affect their interpretation of the geological evidence relating to the so-called "breaks" in the stratigraphic column. For Schindewolf the discontinuities do not indicate an interval of time, however short. It is thus not a simple matter of determining which theory best accounts for the same event. The events which we regard as significant have already been "filled out" or "enriched" in terms of some theory.

The fossil record has not provided a basis for choosing between the two theories, but this is not to say that the arguments presented by Simpson and Schindewolf are without theoretical significance. For several decades Schindewolf carried on, almost single-handed, a vigorous attack upon the widespread view that synthetic theory provided an adequate theoretical foundation for paleontology. Although he gained few adherents to his position, his challenge elicited strong response from paleontologists who subscribed to synthetic theory. The arguments in defense of synthetic theory against the attacks of Schindewolf, particularly those of Simpson, greatly enriched the theoretical content of paleontology. Simpson did not provide compelling support for synthetic theory as against Schindewolfian or Lamarckian, or any number of other theories both evolutionary and nonevolutionary. The claim has been repeatedly made that the fossil record provides a basis for the *falsification* of synthetic theory, and Simpson has demonstrated that this is not the case. He has not shown a way to choose among a number of theories. He has given us good reason not to abandon one of them.

Grene (1958) in a paper that has become famous among paleontologists says of the two theories (p. 185):

> There are, then, two ways of looking at the evolutionary record. The disagreement between the two is not, chiefly, about verifiable or falsifiable matters of fact, but about the concepts through which these facts are to be interpreted. Each constitutes, for its proponent, a closed interpretive system *in* which he sees the facts.

Grene recognized that the fossil record does not provide a basis for choosing between Schindewolfian and synthetic theory, and she sets out to choose between them on other grounds. In contrasting the two theories she says (p. 115):

> To return to Schindewolf: as against the highly abstract and hypothetical Darwinian theory, he makes the assumption: *that life can originate novelty.* This is, he says a preferable assumption because it is simpler. He does not pretend to "explain" this proposition, and in that sense it may be "mysterious," but no more mysterious, he says, than physical concepts like "force" or "gravitation" which everyone is prepared to accept as "explanatory"—

and not nearly so mysterious, he believes, as the whole nexus of assumptions implicit in the concept of "natural selection" as the neo-Darwinians use it.

Classical mechanics, which contains the "mysterious" concepts "force" and "gravity," claims to be comprehensive in the sense that if certain properties and dispositions are attributed to physical objects, then a large number of different kinds of events—including, for example, celestial motion, major overthrusts, and falling bodies—can be explained. Grene seems to be about to say in the passage quoted below (p. 193) that synthetic theory only appears at first sight to be more comprehensive than Schindewolfian theory. She drops this point, however, and goes on to say that synthetic theory overlooks essential aspects of phenomena.

> Simpson's theory, being more complex and abstract, would appear at first sight to cover a wider range of phenomena; yet by the reductive nature of its abstractions it also overlooks essential aspects of the phenomena. Schindewolf's theory is simpler in its reasoning, and less unified in its conceptual structure, but, perhaps for that very reason, it remains closer to the phenomena and does more justice to their experienced complexity.

Darwinian theory claims to be comprehensive in just the sense that mechanics claims to be comprehensive. If certain properties and dispositions are attributed to organisms, it is said, then a great variety of biological events including, according to Simpson, the origin of higher categories in association with major discontinuities in the fossil record can be explained. Those who subscribe to synthetic theory will count an instance of confirmation for a proposition as indirect support for any other proposition to which it is supposed to be logically related in the theory. Thus experiments which support the principles of Mendelian genetics are regarded as indirect support for the claim that higher categories originate by the successive acquisition of characters, just as experiments which support the law of falling bodies are regarded as indirect support for the claim that the planets move about the sun in elliptical orbits.

Schindewolf's evolutionary theory, according to Grene's account of it, rests ultimately upon the claim that *life can originate novelty*, the biological cash value of which seems to be that new major adaptive types originate instantaneously. The cataclysmic origin of new types is the principal explanatory device of the theory. It permits an explanation of the fossil record as adequate as any other. There is, in principle, no reason that theoretical entities should not be major historical events rather than such "mysterious" things as submicroscopic particles and incorporeal motive forces. We can perhaps understand a sense in which a historical event is less mysterious than the force of gravity and closer to the phenomena. But if the principle of the origin of novelty has no consequences other than the sudden appearance of new adaptive types and if it is not itself the consequence of anything else, then its instances of confirmation must come entirely from the fossil record and the not impossible but extremely unlikely event of the origin of a new adaptive type in our own time. Schindewolfian theory, among all evolutionary theories, comes closest to fulfilling Grene's dictum that evolutionary theory be essentially an assessment of the past. This theory, in the opinion of most biologists, provides for an assessment of almost nothing but the past, and is thus less comprehensive than synthetic theory, which claims to provide grounds for an assessment of the past *and* the present.

Now it may be that those biologists who subscribe to synthetic theory are mistaken about its comprehensiveness and thereby mistaken about the amount of evidence which supports it. The claim that a theory is comprehensive is, to an important degree, a logical claim. The attempt to formulate a theory with logical rigor may be viewed as, among other things, an attempt to provide for increase in its potential empirical support. Until Grene, or someone else, demonstrates that synthetic theory is far less comprehensive than its supporters think it is, these supporters will hold that the evidence which can be cited in support of synthetic theory is overwhelmingly greater than that which can be cited in support of Schindewolfian theory.

It is important to note that Schindewolf does not regard his theory of evolution to be as uncomprehensive as Grene makes it out to be, for he cites Goldschmidt (1940) in support of it (see, for example, Schindewolf, 1950, p. 407n). Clearly Goldschmidt and presumably Schindewolf do not regard the origin of new adaptive types to be primitive in the sense that it cannot or need not be explained by subsuming it under more comprehensive hypotheses. Goldschmidt says, for example (p. 395), "He [Schindewolf] shows by examples from fossil material that the major evolutionary advances must have taken place in single large steps, which affected early embryonic stages with the automatic consequence of reconstruction of all later phases of development." Goldschmidt tried throughout his intriguing *The Material Basis of Evolution* (1940), and indeed throughout much of his career, to *explain* major evolutionary advances in single large steps. He was overly optimistic about what Schindewolf or anyone else could show "by examples from fossil material." The fossil record has not provided and will not provide a basis for an evaluation of his genetic and evolutionary theories. These theories are, however, amenable to test by traditional biological methods.

*E*

Recently Eldredge and Gould (1972) have taken up the question of the necessary or contingent discontinuity of the fossil record. They say (p. 98-99):

The idea of *punctuated equilibria* is just as much a preconceived picture as that of phyletic gradualism. We readily admit our bias towards it and urge readers, in the ensuing discussion, to remember that our interpretations are as colored by our preconceptions as are the claims of the champions of phyletic gradualism by theirs. We merely reiterate: (1) that one must have some picture of speciation in mind, (2) that the data of paleontology cannot decide which picture is more adequate, and (3) that the picture of punctuated equilibria is more in accord with the processes of speciation as understood by modern evolutionists.

Eldredge and Gould have taken a conventional view of paleontology. The data of paleontology cannot decide which picture is the more adequate, so it is necessary to go outside to a theoretician, the modern evolutionist, for a judgment. This is a move that Grene is unwilling to make because she does not believe that modern evolutionary theory is sufficiently comprehensive to provide the basis for such a judgment. I suspect also that Grene would regard nonhistorical evidence as irrelevant, because in her view evolutionary theory is essentially an assessment of the past.

Eldredge and Gould do not close the door once and for all on the possibility of the paleontological testing of evolutionary theories, for they say (pp. 108-9):

We have discussed two pictures for the origin of species in paleontology. In the perspective of a species-extrapolation theory of macroevolution, we should now extend these pictures to see how macroevolution proceeds under their guidance. If actual events, as recorded by fossils, fit more comfortably with the predictions of either picture, this will be a further argument for that picture's greater accuracy.

And in an even stronger statement (pp. 93-94), they declare:

Are observed patterns of geographic and stratigraphic distribution, and apparent rates and directions of morphological changes, consistent with the consequences of a particular theory of speciation? We can apply and test, but we cannot generate new mechanisms. If discrepancies are found between paleontological data and the expected patterns, we may be able to identify those aspects of a general theory that need improvement. But we cannot formulate these improvements ourselves.

Why can't paleontologists formulate theoretical improvements themselves? (In asking this question I intend to raise a methodological issue rather than a historical one.) Paleontological events may legitimately be used to test the long-range historical consequences of evolutionary theories if the methodological pitfalls which I have discussed are scrupulously avoided. It is necessary to presuppose a complex theoretical apparatus to get major historical events. It must be remembered, however, that it is necessary to presuppose *some* theoretical apparatus to get any test events whatever. Having identified the aspects of a general theory that need improvement, we need not expect that paleontological events will suggest what the necessary improvement will be. A set of paleontological events cannot provide the basis for the inductive discovery of a new theory. No set of events can provide the basis for the inductive discovery of a theory.

Olson (1960, p. 538) believes that certain problems are posed by the fossil record for selection theory, and among them he lists the following:

> Parallelism in major structures, and particularly in suites of major structures, in evolving lines of populations related only at rather high categorical levels and with remote common ancestors in which the common structures did not exist. Well known cases are found (1) in amphibians, between the major groups apsidospondyls and lepospondyls and within the many groups of apsidospondyls; (2) immense parallelism in the development of multiple similarities in the evolution of the holosteans from the paleoniscoid ancestor; (3) development of suites of mammalian characters in different lines of therapsid reptiles; (4) development of parallel structural features in different lines of ammonites. This seems to be a very prevalent pattern in evolution. It can be explained under selective theory in some cases (for example, Olson, 1959), but it does not seem that its prevalence is something that would be anticipated under the theory in terms of the usual models that are basic to it.

Have events of the kind "parallelism in major structures" occurred? To answer this question requires that immensely complicated historical inferences be performed which presuppose a vast amount of geological and biological knowledge, including the hypothesis that evolution has occurred and some principles of phylogeny construction, but not including the propositions of selection theory. Parallelism in major features may then be considered relevant to the test of selection theory. Olson believes that its prevalence would not be expected under the theory and he is perfectly willing to consider, among other things, the possibility that selection theory may therefore be in need of alternation or enrichment.

It is conceivable in principle and may actually be the case that the reading of the fossil record by Olson and other paleontologists provided a part of the phenomena which pointed to the need of a modification in evolutionary theory to accommodate "punctuated equilibria." Paleontologists did not provide a set of "facts" which compelled any theoretician to provide some particular alteration, but neither did anyone else. As to the claim that "we [paleontologists] cannot formulate these improvements ourselves," Eldredge and Gould may simply be reflecting the reductionist climate which dictates that only genetic answers to theoretical questions are acceptable.

## CONCLUSION

Paleontology provides the means of inferring historical events which can be employed in the testing of evolutionary theories. To accomplish this requires that immensely complicated inferences be performed in which the danger of presupposing the theory to be tested must constantly be guarded against. With the greater use of auxiliary assumptions from geology and with improvements in their own craft, paleontologists will increasingly attempt to draw theoretically significant information from the fossil record.

## REFERENCES CITED

BEERBOWER, J. R., 1968, Search for the past: 2d ed.; Englewood Cliffs, N.J., Prentice-Hall, Inc., 512 p.

CLOUD, P., 1960, Gas as a sedimentary and diagenetic agent: Amer. J. Sci., v. 258A, p. 35-45.

DARWIN, C., 1873, The origin of species by means of natural selection: 6th ed.; London, John Murray, 458 p.

DOBZHANSKY, T., 1956, What is an adaptive trait?: Amer. Natur., v. 90, p. 337-347.

ELDREDGE, N., and GOULD, S. J., 1972, Punctuated equilibria: an alternative to phyletic gradualism, *in* SCHOPF (ed.), Models in paleobiology, San Francisco, Freeman, Cooper and Co., p. 82-115.

GHISELIN, M. T., 1972, Models in phylogeny, *in* SCHOPF (ed.), Models in paleobiology, San Francisco, Freeman, Cooper and Co., p. 130-145.

GOLDSCHMIDT, R., 1940, The material basis of evolution: New Haven, Yale University Press, 436 p.

GOUDGE, T. A., 1961, The ascent of life: Toronto, University of Toronto Press, 236 p.

GRANT, V., 1963, The origin of adaptations: New York, Columbia University Press, 606 p.

GRENE, M., 1958, Two evolutionary theories: Brit. J. Phil. Sci., v. 9, p. 110-127 and 185-193.

________, 1961, Statistics and selection: Brit. J. Phil. Sci., v. 12, p. 25-42.

HARRIS, J. E., 1936, The role of fins in the equilibrium of the swimming fish. I. Wind-tunnel tests on a model of *Mustela canis* (Mitchell): J. Exp. Biol., v. 13, p. 476-493.

JEPSEN, G. L., 1949, Selection, "orthogenesis," and the fossil record: Proc. Amer. Phil. Soc., v. 93, p. 479-500.

LAKATOS, I., 1970, Falsification and the methodology of scientific research programmes, *in* LAKATOS and MUSGRAVE (eds.), Criticism and the growth of knowledge: Cambridge, Cambridge University Press, p. 91-195.

LEWONTIN, R. C., 1957, The adaptations of populations to varying environments: Cold Spring Harbor Sym. Quant. Biol., v. 22, p. 395-408.
OLSON, E. C., 1960, Morphology, paleontology and evolution, *in* TAX (ed.), Evolution after Darwin, v. 1, The evolution of life: Chicago, University of Chicago Press, p. 523-545.
PITTENDRIGH, C. S., 1958, Adaptation, natural selection, and behavior, *in* ROE and SIMPSON (eds.), Behavior and evolution: New Haven, Yale University Press, p. 390-416.
ROMER, A. S., 1966, Vertebrate paleontology: 3rd ed.; Chicago, University of Chicago Press, 468 p.
RUDWICK, M. J. S., 1964, The inference of function from structure in fossils: Brit. J. Phil. Sci., v. 15, p. 27-40.
SCHAEFFER, B., 1965, The role of experimentation in the origin of higher levels of organization: Sys. Zool., v. 14, p. 318-336.
SCHINDEWOLF, O. H., 1950, Grunfragen der Paläontologie: Stuttgart, E. Schweizerbart, 505 p.
SCHOPF, T. J. M., 1972, About this book, *in* SCHOPF (ed.), Models in paleobiology: San Francisco, Freeman, Cooper, and Co., 237 p.
SIMPSON, G. G., 1944, Tempo and mode in evolution: New York, Columbia University Press.
———, 1953, The major features of evolution: New York, Columbia University Press, 434 p.

# Notes

## CHAPTER ONE

1. When this paper was written I accepted the view, common among geologists, that the lack of deductive rigor in geological inference could best be treated as probabilistic in character. I have discussed a modified view of uncertainty in geology in Chapter Two.

2. It is well to emphasize here the distinction between discovery and justification. When a geologist is told that historical inference may be viewed within the context of some theory of explanation, it may appear to him that he is being told that in discovering some geological event he has constructed an explanation according to the canons of that theory. The geologist will almost certainly deny that he has done any such thing, for the very good reason that he hasn't. If, on the other hand, a geologist is asked to *justify* his contention that some event has occurred, he may well produce an *argument* which turns out to have many features explicitly identified by some theory of explanation.

3. There is no question that geologists generalize from instances which they have observed. But, as I have pointed out in the introduction, those instances are always described in terms already imbued with theoretical significance; and furthermore, the process of generalization itself proceeds within the context of theoretical preconceptions. Given these indisputable facts, there seems to be less of a role for induction in geological discovery than most geological methodologists have supposed.

4. The problem of the lack of independent support for the antecedent statements in geological inference is discussed in detail in Chapter Two.

5. I used *prediction* too loosely here. A prediction is an inference in which a statement about the future is derived from a statement about the present, or the past, together with some generalizations. Neither Swineford nor Hutton made a *geological* prediction, which is to say that they did not make a statement about some future state of the earth. They based a rational expectation of what they might find upon geological evidence which was wholly retrodictive in its foundation. When geologists claim that their science is predictive they often point to cases of this kind. Whatever the precise nature of this inference, it assumes a large role in economic geology.

## CHAPTER TWO

1. One of the exceptions is the *pre-indication* produced by a scientist in the form of a picture or descriptive statement when he makes a prediction.

## CHAPTER THREE

1. My present position on probability in geology is contained in Chapter Two.

2. The use of *deductive* is unclear here, as was the use of *inductive* in a previous instance. When I say that advances will be deductive I am not pointing to an increasing rigor, but to an increasing tendency to formulate geological generalizations in conscious and explicit relationship to physical theories which are supposed to comprehend them. This tendency is what geologists are pointing to when they claim that geology is becoming less "descriptive" and more "scientific."

3. I have come to regard my treatment of the uniformitarian principle in this paper as pedantic and laborious, although not wrong. I am pleased to think that Gould said essentially the same thing about uniformity, but with more clarity and elegance, in his excellent paper of 1965.

## CHAPTER FOUR

1. My colleague Sabetai Unguru has pointed out to me that in this passage I might be taken to be claiming that the law of superposition permits temporal ordering without any theoretical preconceptions whatever. I certainly do not mean to claim this, nor is it necessary to do so in order to support my argument. It is only necessary to show that on occasion we can order two events by superposition without first having ordered them on the hypothesis that they are causally related.

## CHAPTER FIVE

1. Gilbert's example for the "inculcation of the scientific method" is his ingenious attempt to explain the differences in elevation of the shoreline of the Pleistocene Lake Bonneville.

2. *Induction* and *generalization* are frequently used by geologists, and sometimes by historians, to describe an argument that leads to conclusions which are "broad" rather than general, where "broad" stands for wide spatial and temporal extent. Thus Turner's Frontier Hypothesis is sometimes cited by historians as a historical generalization.

## CHAPTER SIX

A draft of this paper was written while I was on sabbatical leave at Stanford University in the spring of 1972; it was presented to the geology colloquium there. I extend my thanks to Norman J. Silberling, with whom I discussed many of the problems treated here.

## CHAPTER SEVEN

This chapter is a revised version of a paper published in 1966 in the *Journal of Geology* and cited elsewhere.

## CHAPTER EIGHT

1. Some paleontologists believe that this theoretical foundation has already been provided. In response to my comments (Kitts 1975) on their paper on phylogeny (1973), Raup et al. (1975) reply,

> We cannot, however, accept Kitts' suggestion that the principles of constructing phylogenies have not been explicitly stated, or are kept secret ("covertly formulated") by practitioners. The literature of phylogenetic procedure and methodology is large and constantly growing—especially in response to a host of new and old problems raised by work in the field of numerical taxonomy.

The admittedly large literature on phylogenetic procedure has not, in my opinion, succeeded in elucidating the method of paleontological phylogeny construction, nor do I believe that it has had much impact upon paleontologists. More important, however, is the fact that the phylogeny of reptiles which Raup et al. adduced for comparative purposes was taken from Harland et al. (1967), who had, in turn, relied upon a number of different authorities. There is no evidence that any of these authorities had been influenced by the current literature on phylogeny construction. There is, in fact, good reason to think that they operated in an older tradition, the theoretical foundations of which have hardly been considered. Unless we suppose that the recent work on phylogeny has succeeded in explicating the principles that paleontological phylogeny constructors invoke, then we are still left without what we need most to judge the treatment of Raup et al., and that is an explicit statement of the theoretical and methodological foundations of the paleontological phylogeny which figured so importantly in that treatment.

## References in Notes

Gould, S. J., 1965, Is uniformitarianism necessary: Am. Jour. Sci., v. 263, p. 223-228.

Harland, W. B., et al., 1967, The fossil record: London, Geological Society, 828 p.

Kitts, D. B., 1975, Stochastic models of phylogeny and the evolution of diversity: a discussion: Jour. Geology, v. 83, p. 125-126.

Raup, D. M.; Gould, S. J.; Schopf, T. J. M.; and Simberloff, D. S., 1973, Stochastic models of phylogeny and the evolution of diversity: Jour. Geology, v. 81, p. 525-542.

———, 1975, Stochastic models of phylogeny and the evolution of diversity: a reply: Jour. Geology, v. 83, p. 126-127.

# Index

THE STRUCTURE OF GEOLOGY is an account of the structure of geological knowledge that examines the complex inferential context in which statements about the past are derived and tested. Emphasized is the dependence of virtually all geological inferences upon an unquestioned background of contemporary physical theory. A theory is proposed in which the concept of the signal, so familiar in contemporary physics, is introduced in an attempt to solve some long-standing difficulties concerning geologic time. The widely held view that geology has recently passed through a scientific revolution is carefully examined and the opinion is offered that the dramatic change in earth science during the past twenty years, although in many ways like the scientific revolutions described by Thomas Kuhn, has unique features that grow out of the historical character of geology. The relationship between paleontology and evolutionary theory is also considered.

DAVID B. KITTS holds the A.B. degree from the University of Pennsylvania and the Ph.D. from Columbia University. He is presently David Ross Boyd Professor of Geology and Geophysics and of the History of Science, and chairman of the Department of the History of Science at the University of Oklahoma.

SOUTHERN METHODIST UNIVERSITY PRESS • DALLAS